Lectures in Mathematics
ETH Zürich
Department of Mathematics
Research Institute of Mathematics

Managing Editor:
Michael Struwe

Leonid Polterovich
The Geometry of the Group of Symplectic Diffeomorphism

Birkhäuser Verlag
Basel · Boston · Berlin

Author's address:

School of Mathematical Sciences
Tel Aviv University
69978 Tel Aviv
Israel

2000 Mathematical Subject Classification 53Dxx, 57R17, 58Dxx, 37Jxx, 70Hxx

A CIP catalogue record for this book is available from the
Library of Congress, Washington D.C., USA

Deutsche Bibliothek Cataloging-in-Publication Data
Polterovich, Leonid:
The geometry of the group of symplectic diffeomorphisms / Leonid Polterovich. -
Basel ; Boston ; Berlin : Birkhäuser, 2001
   (Lectures in mathematics)
   ISBN 3-7643-6432-7

ISBN 3-7643-6432-7 Birkhäuser Verlag, Basel – Boston – Berlin

© 2001 Birkhäuser Verlag, P.O. Box 133, CH-4010 Basel, Switzerland
Member of the BertelsmannSpringer Publishing Group
Printed on acid-free paper produced from chlorine-free pulp. TCF ∞
Printed in Germany
ISBN 3-7643-6432-7

9 8 7 6 5 4 3 2 1

Contents

7 Diameter

8 Growth and Dynamics

9 Length Spectrum

10 Deformations of Symplectic Forms

11 Ergodic Theory

12 Geodesics

13 Floer Homology

14 Non-Hamiltonian Diffeomorphisms

Preface

The group of Hamiltonian diffeomorphisms $\mathrm{Ham}(M,\Omega)$ of a symplectic manifold (M,Ω) plays a fundamental role both in geometry and classical mechanics. For a geometer, at least under some assumptions on the manifold M, this is just the connected component of the identity in the group of all symplectic diffeomorphisms. From the viewpoint of mechanics, $\mathrm{Ham}(M,\Omega)$ is the group of all admissible motions. What is the minimal amount of energy required in order to generate a given Hamiltonian diffeomorphism f? An attempt to formalize and answer this natural question has led H. Hofer [H1] (1990) to a remarkable discovery. It turns out that the solution of this variational problem can be interpreted as a *geometric quantity*, namely as the distance between f and the identity transformation. Moreover this distance is associated to a canonical biinvariant metric on $\mathrm{Ham}(M,\Omega)$. Since Hofer's work this new geometry has been intensively studied in the framework of modern symplectic topology. In the present book I will describe some of these developments.

Hofer's geometry enables us to study various notions and problems which come from the familiar finite dimensional geometry in the context of the group of Hamiltonian diffeomorphisms. They turn out to be very different from the usual circle of problems considered in symplectic topology and thus extend significantly our vision of the symplectic world. Is the diameter of $\mathrm{Ham}(M,\Omega)$ finite or infinite? What are the minimal geodesics? How can one find the length spectrum? In general, these questions are still open. However some partial answers do exist and will be discussed below.

There is one more, to my taste even more important reason why it is useful to have a canonical geometry on the group of Hamiltonian diffeomorphisms. Consider a time dependent vector field $\xi_t, t \in \mathbb{R}$ on a manifold M. The ordinary differential equation

$$\dot{x} = \xi(x,t)$$

on M defines a flow $f_t : M \to M$ which takes any initial condition $x(0)$ to the solution $x(t)$ at time t. The trajectories of the flow form a complicated system of curves on the manifold. Usually, in order to understand the dynamics, one should travel along the manifold and thoroughly study behaviour of the trajectories in different regions. Let us change the point of view and note that our flow can be interpreted as a simple geometric object – a single curve $t \to f_t$ on the group of all diffeomorphisms of the manifold. One may hope that geometric properties of this curve reflect the dynamics, and thus complicated dynamical phenomena can be studied by purely geometric tools. Of course, the price we pay is that the ambient space – the group of diffeomorphisms – is infinite-dimensional. Moreover, we face a serious difficulty: in general there are no canonical tools to perform geometric measurements on this group. Remarkably, Hofer's metric provides such a tool for systems of classical mechanics. In Chapters 8 and 11 we will see some situations where this ideology turns out to be successful.

As it often happens with very young and fast developing areas of mathematics, the proofs of various elegant statements in Hofer's geometry are technical and complicated. Thus I selected the simplest non-trivial versions of the main phenomena (according to my taste, of course) without making an attempt to present them in the most general form. For the same reason many technicalities are omitted. Though formally speaking this book does not require a special background in symplectic topology (at least necessary definitions and formulations are given), the reader is cordially invited to consult two remarkable introductory texts [HZ] and [MS]. Both of them contain chapters on the geometry of the group of Hamiltonian diffeomorphisms. I have tried to minimize the overlaps. The book contains a number of exercises which presumably will help the reader to get into the subject.

This book arose from two sources. The first one is lectures given to graduate students, namely mini-courses at Universities of Freiburg and Warwick, and a *Nachdiplom*-course at ETH, Zürich. The second source is the expository article [P8] which contains a brief outline of the material presented below.

Let me introduce briefly the main characters of the book. A diffeomorphism f of a symplectic manifold (M, Ω) is called *Hamiltonian* if it can be included into a compactly supported Hamiltonian flow $\{f_t\}$ with $f_0 = \mathbb{1}$ and $f_1 = f$. Such a flow is defined by a Hamiltonian function $F : M \times [0; 1] \to \mathbb{R}$. In the language of classical mechanics F is the energy of a mechanical motion described by $\{f_t\}$. We interpret the total energy of the flow as a length of the corresponding path of diffeomorphisms:

$$\text{length}\{f_t\} = \int_0^1 \max_{x \in M} F(x, t) - \min_{x \in M} F(x, t) dt.$$

Define a function

$$\rho : \text{Ham}(M, \Omega) \times \text{Ham}(M, \Omega) \to \mathbb{R}$$

by

$$\rho(\phi, \psi) = \inf \text{length}\{f_t\},$$

where the infimum is taken over all Hamiltonian flows $\{f_t\}$ which generate the Hamiltonian diffeomorphism $f = \phi\psi^{-1}$. It is easy to see that ρ is non-negative, symmetric, vanishes on the diagonal and satisfies the triangle inequality. Moreover, ρ is biinvariant with respect to the group structure of $\mathrm{Ham}(M,\Omega)$. In other words, ρ is a biinvariant pseudo-distance. It is a deep fact that ρ is a genuine distance function, that is $\rho(\phi,\psi)$ is strictly positive for $\phi \neq \psi$. The metric ρ is called Hofer's metric.

The group $\mathrm{Ham}(M,\Omega)$ and Hofer's pseudo-distance ρ are introduced in Chapters 1 and 2 respectively. In Chapter 3 we prove that ρ is a genuine metric in the case when M is the standard symplectic linear space $\mathbb{R}^{2n}$. Our approach is based on Gromov's theory of holomorphic discs with Lagrangian boundary conditions which is presented in Chapter 4.

Afterwards we turn to the study of basic geometric invariants of $\mathrm{Ham}(M,\Omega)$. There is a (still open!) conjecture that the diameter of $\mathrm{Ham}(M,\Omega)$ is infinite. In Chapters 5–7 we prove this conjecture for closed surfaces.

In Chapter 8 we discuss the concept of the growth of a one-parameter subgroup $\{f_t\}$ of $\mathrm{Ham}(M,\Omega)$ which reflects the asymptotic behaviour of the function $\rho(\mathbb{1}, f_t)$ when $t \to \infty$. We present a link between the growth and the dynamics of $\{f_t\}$ in the context of invariant tori of classical mechanics.

In many interesting situations the space $\mathrm{Ham}(M,\Omega)$ has complicated topology, and in particular a non-trivial fundamental group. For an element $\gamma \in \pi_1(\mathrm{Ham}(M,\Omega))$ set $\nu(\gamma) = \inf \mathrm{length}\{f_t\}$, where the infimum is taken over all loops $\{f_t\}$ of Hamiltonian diffeomorphisms (that is periodic Hamiltonian flows) which represent γ. The set

$$\{\nu(\gamma) \mid \gamma \in \pi_1(\mathrm{Ham}(M,\Omega))\}$$

is called the length spectrum of $\mathrm{Ham}(M,\Omega)$. In Chapter 9 we present an approach to the length spectrum which is based on the theory of symplectic fibrations. An important ingredient of this approach is Gromov's theory of pseudo-holomorphic curves which is discussed in Chapter 10. In Chapter 11 we give an application of our results on the length spectrum to the classical ergodic theory.

In Chapters 12 and 13 we develop two different approaches to the theory of geodesics on $\mathrm{Ham}(M,\Omega)$. One of them is elementary, while the other requires a powerful machinery of Floer homology. Chapter 13 suggests to the reader a brief visit to Floer homology.

Finally, in Chapter 14 we deal with non-Hamiltonian symplectic diffeomorphisms, which appear naturally in Hofer's geometry as isometries of $\mathrm{Ham}(M,\Omega)$. In addition, we formulate and discuss the famous flux conjecture which states that the group $\mathrm{Ham}(M,\Omega)$ is closed in the group of all symplectic diffeomorphisms endowed with the C^∞-topology.

Acknowledgments

I cordially thank Meike Akveld for her indispensable help in typing the preliminary version of the manuscript, preparing the pictures and enormous editorial work. I am very grateful to Paul Biran and Karl Friedrich Siburg for their detailed comments on the manuscript and for improving the presentation. I am indebted to Rami Aizenbud, Dima Gourevitch, Misha Entov, Osya Polterovich and Zeev Rudnick for pointing out a number of inaccuracies in the preliminary version of the book. The book was written during my stay at ETH, Zürich in the academic year 1997–1998, and during my visits to IHES, Bures-sur-Yvette in 1998 and 1999. I thank both institutions for the excellent research atmosphere.

Chapter 1

Introducing the Group

In this chapter we sum up some well-known preliminary facts about the group of Hamiltonian diffeomorphisms.

1.1 The origins of Hamiltonian diffeomorphisms

Consider the motion of a mass 1 particle in $\mathbb{R}^n(q)$, where q denotes the coordinate on $\mathbb{R}^n$, in the presence of a potential force $\Phi(q,t) = -\frac{\partial U}{\partial q}(q,t)$. Newton's second law yields $\ddot{q} = \Phi(q,t)$. In general there is no chance to solve this equation explicitly except for some rare examples. However it is possible to understand some qualitative properties.

Let us make a little trick. Introduce an auxiliary variable $p = \dot{q}$, and consider the function $F(p,q,t) = \frac{p^2}{2} + U(q,t)$. The function F represents the total energy of the particle (the sum of the kinetic and the potential energies). With this notation the Newton equation above can be rewritten as follows:

$$\begin{cases} \dot{p} = -\frac{\partial F}{\partial q}(p,q,t) \\[2mm] \dot{q} = \frac{\partial F}{\partial p}(p,q,t). \end{cases}$$

This system of the first order differential equations is called *Hamiltonian system*. It should be considered as a differential equation in the $2n$-dimensional space $\mathbb{R}^{2n}$ endowed with coordinates p and q. The first step of any qualitative study is to forget the explicit form of the object you are interested in. With this ideology at hand let us leave aside the expression for the function F above and focus on properties of general Hamiltonian systems associated with more or less arbitrary smooth energy functions $F(p,q,t)$. "More or less" means that we impose certain assumptions on the behaviour of F at infinity in order to guarantee that the solutions of the Hamiltonian system exist for all $t \in \mathbb{R}$. So choose such an F and consider the flow $f_t : \mathbb{R}^{2n} \to \mathbb{R}^{2n}$ which takes any initial condition $(p(0), q(0))$ to the corresponding solution $(p(t), q(t))$ at time t. Informally, we call diffeomorphisms f_t arising in this way *mechanical motions*. The essence of our trick is in the fact that these diffeomorphisms of the $2n$-dimensional space $\mathbb{R}^{2n}$ have the following remarkable geometric properties which cannot be seen in the original configuration space $\mathbb{R}^n$.

Theorem 1.1.A. *(Liouville Theorem) Mechanical motions preserve the volume form* $\mathrm{Vol} = dp_1 \wedge dq_1 \wedge \cdots \wedge dp_n \wedge dq_n$.

Theorem 1.1.B. *(A more delicate variant of 1.1.A) Mechanical motions preserve the 2-form* $\omega = dp_1 \wedge dq_1 + \cdots + dp_n \wedge dq_n$.

Note that $\mathrm{Vol} = \frac{\omega^n}{n!}$ and hence 1.1.B implies 1.1.A. Further, for $n = 1$, Theorems 1.1.A and 1.1.B are equivalent. Theorem 1.1.B is a simple consequence of the fact that mechanical motions come from Hamiltonian system. We leave the proof till the next section.

Both results above go back to the past. The preservation of volume by mechanical motions attracted a lot of attention already more than a century ago. It served as the main stimulating force for creation of *ergodic theory*, nowadays a well established mathematical discipline which studies various recurrence properties of measure preserving transformations. However the significance of the role played by the invariant 2-form ω has been noticed relatively recently. As far as I know it was V.I. Arnold who first pointed it out explicitly in the 1960s. An attempt to understand the difference between mechanical motions and general volume preserving diffeomorphisms gave rise to the field of *symplectic topology* which investigates surprising rigidity phenomena appearing in the theory of symplectic manifolds and their morphisms.

Here is an example of such a phenomenon which was found by J.-C. Sikorav in [S1]. Let $B^2(r) \subset \mathbb{R}^2$ be the Euclidean disc of radius r bounded by the circle $S^1(r)$. Consider a torus

$$L_R = S^1(R) \times \cdots \times S^1(R) \subset \mathbb{R}^2(p_1, q_1) \times \cdots \times \mathbb{R}^2(p_n, q_n) = \mathbb{R}^{2n}(p, q)$$

and a cylinder $C_r = B^2(r) \times \mathbb{R}^{2n-2}$.

Theorem 1.1.C. *(A non-squeezing property) There is no mechanical motion taking* L_R *into* C_r *provided* $R > r$.

We will prove this statement in a more general context in section 3.2.E below. Note that for $n = 1$ the result is obvious. Indeed, the area bounded by $S^1(R)$ is greater than the area of $B^2(r)$. Thus one cannot take $S^1(R)$ into $B^2(r)$ by an area preserving transformation. However, if $n \geq 2$ then L_R is a submanifold of codimension $n \geq 2$, while C_r has infinite volume. So there is no visible reason why the statement should hold true. Moreover, it is certainly wrong in the volume-preserving category!

Exercise. Find a linear volume preserving transformation $\mathbb{R}^{2n} \to \mathbb{R}^{2n}$ taking L_R into C_r for arbitrary positive r and R.

In what follows we shall consider the evolution of a mechanical system as a curve in the group of all mechanical motions and study this curve by geometric tools. To make things work we are forced to restrict the class of mechanical systems we deal with. For instance, unbounded Hamiltonians such as $F(p, q, t) = \frac{p^2}{2} + U(q, t)$ above are too complicated to be included into such a framework. In fact we will always assume that Hamiltonian functions (and therefore the corresponding mechanical motions) are compactly supported. In other words all the action takes place in a bounded part of the space we live in.

In this chapter we introduce Hamiltonian mechanics on symplectic manifolds which is a natural generalization of the model described above. In this context a Hamiltonian diffeomorphism is just a mechanical motion generated by a compactly supported Hamiltonian function.

1.2 Flows and paths of diffeomorphisms

To start with, we explain the relation between flows and differential equations, and give a geometric interpretation of flows as paths of diffeomorphisms. In view of future applications, we will assume that all objects we consider are compactly supported. However, basic concepts described below can be extended in a straightforward way to more general situations.

Consider a smooth manifold M without boundary. For a diffeomorphism $\phi : M \to M$ define its support $\mathrm{supp}(\phi)$ as the closure of all $x \in M$ such that $\phi x \neq x$. Denote by $\mathrm{Diff}^c(M)$ the group of all diffeomorphisms with compact support. Let $I \subset \mathbb{R}$ be an interval.[1] A *path* of diffeomorphisms is a map

$$f : I \to \mathrm{Diff}^c(M), \ t \to f_t$$

with the following properties:

- the map $M \times I \to M$ taking (x, t) to $f_t x$ is smooth;
- there exists a compact subset K of M which contains $\mathrm{supp}\, f_t$ for all $t \in I$.

We often denote such a path by $\{f_t\}$. Note that on closed manifolds the second condition is automatically satisfied.

---

[1]We define an interval as a connected subset of $\mathbb{R}$ with non-empty interior.

Every path of diffeomorphisms generates a family of vector fields ξ_t, $t \in I$ on M as follows:

$$\frac{d}{dt} f_t x = \xi_t(f_t x). \qquad (1.2.A)$$

Note that this family is smooth and compactly supported: $\xi_t(x) = 0$ for all $x \in M \setminus K$. Such a family is called a *compactly supported time-dependent vector field* on M. The correspondence above is not injective. Indeed, every path of the form $\{f_t g\}$ where g is an arbitrary element of $\mathrm{Diff}^c(M)$ generates the same time-dependent vector field ξ. However, for every point $s \in I$ there exists a unique path $\{f_t\}$ which generates ξ such that f_s equals the identity map $\mathbb{1}$. This path is defined as the unique solution of 1.2.A which is considered now as an ordinary differential equation with the initial condition $f_s = \mathbb{1}$. Assume that $0 \in I$ and take $s = 0$. The path $\{f_t\}$ constructed above with $f_0 = \mathbb{1}$ is called a *flow* of the time-dependent vector field ξ. So flows are just paths $\{f_t\}$ satisfying $f_0 = \mathbb{1}$.

1.3   The mathematical model of classical mechanics

The role of the phase space in classical mechanics is played by a symplectic manifold (M^{2n}, Ω). Here M is a connected manifold without boundary of even dimension $2n$, and Ω is a closed differential 2-form on M. The form Ω is assumed to be *non-degenerate*. This means that its top power Ω^n does not vanish at any point. The form $\mathrm{Vol} = \frac{1}{n!}\Omega^n$ is called *the canonical volume form* on (M, Ω). It is useful to keep in mind two elementary examples of symplectic manifolds: an orientable surface endowed with an area form, and the linear space $\mathbb{R}^{2n}(p_1, \ldots, p_n, q_1, \ldots, q_n)$ with the form $\omega = \Sigma_{j=1}^n dp_j \wedge dq_j$. The second example is of great importance in view of the classical Darboux theorem [MS]. It states that locally every symplectic manifold looks like $(\mathbb{R}^{2n}, \omega)$. In other words, near each point of M one can choose local coordinates (p, q) such that in these coordinates Ω is written as $\Sigma_{j=1}^n dp_j \wedge dq_j$. We call (p, q) *canonical local coordinates*.

Let F be a smooth function on M. A vector field ξ on M is called *the Hamiltonian vector field* of F if it satisfies the pointwise linear algebraic equation $i_\xi \Omega = -dF$. An elementary argument from linear algebra (based on the non-degeneracy of Ω) shows that ξ always exists and is unique [MS]. Sometimes ξ is denoted by $\mathrm{sgrad}\, F$ (skew gradient of F).

Exercise 1.3.A. Show that in canonical local coordinates (p, q) on M one has $\mathrm{sgrad}\, F = (-\frac{\partial F}{\partial q}, \frac{\partial F}{\partial p})$.

Exercise 1.3.B. Let $\phi : M \to M$ be a *symplectic diffeomorphism* (that is $\phi^*\Omega = \Omega$). Show that

$$\mathrm{sgrad}(F \circ \phi^{-1}) = \phi_* \, \mathrm{sgrad} \, F$$

for every function F on M. This property, of course, reflects the fact that the operation sgrad is defined in the coordinate-free manner.

In classical mechanics energy determines the evolution. The energy is a family of functions F_t on M which depends on the additional time variable t. The time t is defined on some interval I. Equivalently, one can consider the energy as a single function F on $M \times I$. We shall use both points of view throughout the book, and keep the notation $F_t(x) = F(x, t)$. Traditionally, F is called a *(time-dependent) Hamiltonian function*.

The evolution of the system is described by the *Hamiltonian equation* $\dot{x} = \mathrm{sgrad}\, F_t(x)$. In local canonical coordinates (p, q) on M the Hamiltonian equation has the familiar form (cf. 1.3.A)

$$\begin{cases} \dot{p} = -\frac{\partial F}{\partial q}(p, q, t) \\[2mm] \dot{q} = \frac{\partial F}{\partial p}(p, q, t). \end{cases}$$

Let us introduce a linear function space $\mathcal{A} = \mathcal{A}(M)$ which plays an important role below. When M is closed, define $\mathcal{A}(M)$ as the space of all smooth functions on M with zero mean with respect to the canonical volume form. When M is open, $\mathcal{A}(M)$ consists of all smooth functions with compact support.

Definition. Let $I \subset \mathbb{R}$ be an interval. A (time-dependent) Hamiltonian function F on $M \times I$ is called **normalized** if F_t belongs to $\mathcal{A}$ for all t. In the case when M is open we require in addition that there exists a compact subset of M which contains the supports of all the functions F_t, $t \in I$ simultaneously.

In what follows we consider normalized Hamiltonians only. Let me give some arguments in favor of this decision.

First of all, on open manifolds one must impose some assumptions on the behaviour of Hamiltonian functions at infinity. Indeed, otherwise the solutions of the Hamiltonian equation may blow up at finite time and thus the Hamiltonian flow may be ill defined. An important feature of the definition above is that the time dependent Hamiltonian vector field $\mathrm{sgrad}\, F_t$ of a normalized Hamiltonian function F is compactly supported. Thus when I contains 0 such a field generates a compactly supported flow which fits into the setting of the previous section.

Second, both on open and closed manifolds, the mapping which takes a function from $\mathcal{A}$ to its Hamiltonian vector field is injective. Indeed, a Hamiltonian vector field determines the corresponding Hamiltonian function uniquely up to an additive constant. Clearly, our normalization forbids to add constants! This property of normalized Hamiltonian functions will be useful for our further considerations.

1.4   The group of Hamiltonian diffeomorphisms

Let $F : M \times I \to \mathbb{R}$ be a normalized time dependent Hamiltonian function. Assume that I contains zero. Consider the flow $\{f_t\}$ of the time-dependent vector field sgrad F_t. We will say that $\{f_t\}$ is *the Hamiltonian flow* generated by F. Every individual diffeomorphism f_a, $a \in I$ of this flow is called *a Hamiltonian diffeomorphism*. Of course, the definition implies that Hamiltonian diffeomorphisms are compactly supported.

Exercise 1.4.A (reparametrization of flows) Let $\{f_t\}$, $t \in [0; a]$ be a Hamiltonian flow generated by a normalized Hamiltonian $F(x, t)$. Show that $\{f_{at}\}$, $t \in [0; 1]$ is again a Hamiltonian flow generated by $aF(x, at)$. Therefore every Hamiltonian diffeomorphism is in fact a time-one map of some Hamiltonian flow. More generally, show that for every smooth function $b(t)$ with $b(0) = 0$ the flow $\{f_{b(t)}\}$ is a Hamiltonian flow whose normalized Hamiltonian equals $\frac{db}{dt}(t)F(x, b(t))$.

The crucial property of Hamiltonian diffeomorphisms is that they preserve the symplectic form Ω. Indeed let ξ be the Hamiltonian vector field of a function F on M. All we have to show is that the Lie derivative $\mathcal{L}_\xi \Omega$ vanishes. This can be seen from the following calculation:

$$\mathcal{L}_\xi \Omega = i_\xi d\Omega + d(i_\xi \Omega) = -ddF = 0.$$

Denote by $\mathrm{Ham}(M, \Omega)$ the set of all Hamiltonian diffeomorphisms.

A path of diffeomorphisms with values in $\mathrm{Ham}(M, \Omega)$ is called a *Hamiltonian path*. It sounds natural to call a flow with values in $\mathrm{Ham}(M, \Omega)$ a Hamiltonian flow. However, after a little thinking, we realize that we are in trouble. Indeed, Hamiltonian flows were already defined above in a different way. A priori it is not at all clear why the vector field corresponding to a flow of Hamiltonian diffeomorphisms is a Hamiltonian vector field! Fortunately, this is true. This extremely important fact was established by Banyaga in [B1]. Let us formulate it precisely.

Proposition 1.4.B. *For every Hamiltonian path $\{f_t\}$, $t \in I$ there exists a (time-dependent) normalized Hamiltonian function $F : M \times I \to \mathbb{R}$ such that*

$$\frac{d}{dt} f_t x = \mathrm{sgrad}\, F_t(f_t x)$$

for all $x \in M$ and $t \in I$.

The function F is called the normalized Hamiltonian function of $\{f_t\}$.

Let us discuss this result. First assume that the manifold M satisfies the following topological condition: its first de Rham cohomology with compact supports vanishes, that is $H^1_{\text{comp}}(M, \mathbb{R}) = 0$. To have an example in mind, think of the 2-sphere or the linear space. In this case the proposition above can be proved very easily. Denote by ξ_t the vector field generated by f_t. Since Hamiltonian diffeomorphisms preserve the symplectic form Ω we have $\mathcal{L}_{\xi_t}\Omega = 0$. Hence $di_{\xi_t}\Omega = 0$ so $i_{\xi_t}\Omega$ is a closed form. In view of our topological condition the form $i_{\xi_t}\Omega$ is exact. Therefore there exists unique smooth family of functions $F_t(x) \in \mathcal{A}$ such that $-dF_t = i_{\xi_t}\Omega$. It follows that $F(x,t)$ is the normalized Hamiltonian of $\{f_t\}$. This completes the proof. However when $H^1_{\text{comp}}(M, \mathbb{R}) \neq 0$, we do not know a priori that closed forms $i_{\xi_t}\Omega$ are exact. In this case one should use in addition that each individual f_t is Hamiltonian. This requires some new ideas, see [B1], [MS].

Remark 1.4.C. Let $\text{Symp}(M, \Omega)$ be the group of all compactly supported diffeomorphisms f of M which preserve Ω, that is $f^*\Omega = \Omega$. Such diffeomorphisms are called *symplectomorphisms*. Denote by $\text{Symp}_0(M, \Omega)$ the path connected component of the identity in $\text{Symp}(M, \Omega)$. By definition, it contains those symplectomorphisms f which can be joined with the identity by a path of symplectomorphisms. Arguing exactly as in the proof of 1.4.B above we see that such a path is Hamiltonian provided $H^1_{\text{comp}}(M, \mathbb{R}) = 0$. Therefore in this case

$$\text{Symp}_0(M, \Omega) = \text{Ham}(M, \Omega).$$

When $H^1_{\text{comp}}(M, \mathbb{R}) \neq 0$ this equality is not valid anymore. For instance, consider a 2-dimensional torus $T^2 = \mathbb{R}^2(p, q)/\mathbb{Z}^2$ endowed with the area form $\Omega = dp \wedge dq$. A translation $(p, q) \to (p + a, q)$ obviously lies in $\text{Symp}_0(T^2, \Omega)$. However, for non-integer a one can show that it is not Hamiltonian. On the other hand the difference between Ham and Symp_0 is not "too big" and it can be described in a rather simple way, see 14.1 below.

The next proposition contains a remarkable elementary formula which will be used below on many occasions.

Proposition 1.4.D. *(Hamiltonian of the product) Consider two Hamiltonian paths $\{f_t\}$ and $\{g_t\}$. Let F and G be their normalized Hamiltonian functions. Then the product path $h_t = f_t g_t$ is a Hamiltonian path generated by the normalized Hamiltonian function*

$$H(x, t) = F(x, t) + G(f_t^{-1}x, t).$$

Let us prove this formula. We are given that

$$\frac{d}{dt}f_t x = \text{sgrad}\, F_t \quad \text{and} \quad \frac{d}{dt}g_t x = \text{sgrad}\, G_t.$$

Thus

$$\frac{d}{dt}(f_t g_t)x = \operatorname{sgrad} F_t + f_{t*} \operatorname{sgrad} G_t.$$

The second summand in the right-hand side equals $\operatorname{sgrad}(G \circ f_t^{-1})$ in view of Exercise 1.3.B above. We conclude that

$$\frac{d}{dt}h_t x = \operatorname{sgrad}(F_t + G \circ f_t^{-1}) = \operatorname{sgrad} H_t.$$

The formula is proved.

Now we are ready to justify the title of this section.

Proposition 1.4.E. *The set of Hamiltonian diffeomorphisms is a group with respect to composition.*

Indeed, take two Hamiltonian diffeomorphisms f and g. In view of 1.4.A we can write $f = f_1$ and $g = g_1$ for some Hamiltonian flows $\{f_t\}, \{g_t\}$ defined for $t \in [0;1]$. Proposition 1.4.D yields that that the path $\{f_t g_t\}$ is a Hamiltonian flow. Thus its time-one map fg is a Hamiltonian diffeomorphism. So the set $\operatorname{Ham}(M, \Omega)$ is closed under composition of diffeomorphisms. It remains to verify that f^{-1} is a Hamiltonian diffeomorphism. This follows from the next exercise.

Exercise. Show that the path $\{f_t^{-1}\}$ is a Hamiltonian flow generated by Hamiltonian $-F(f_t x, t)$. *Hint:* differentiate the identity $f_t \circ f_t^{-1} = \mathbb{1}$ with respect to t and argue as in the proof of 1.4.D above.

Remark 1.4.F. In differential geometry one usually deals with groups of transformations which preserve certain structures on a manifold (such as the group $\operatorname{Symp}(M, \Omega)$ of all symplectomorphisms). The group of Hamiltonian diffeomorphisms does not have such a friendly description (Hamiltonian diffeomorphisms are not defined as morphisms in certain natural category). This leads to very surprising complications. For instance the next question (known as the Flux conjecture, see Chapter 14) is still open for most symplectic manifolds M. Suppose that M is closed. Assume that some sequence of Hamiltonian diffeomorphisms C^∞-converges to a symplectic diffeomorphism f. Is f Hamiltonian?

It is extremely useful to think about $\operatorname{Ham}(M, \Omega)$ in terms of Lie group theory. This point of view is fundamental for a development of geometric intuition which is necessary for our purposes. So let us work out this language. We will consider $\operatorname{Ham}(M, \Omega)$ as a Lie subgroup of the group of all diffeomorphisms of M. Thus the Lie algebra[2] of $\operatorname{Ham}(M, \Omega)$ is just the algebra of all

---

[2]As a vector space, the Lie algebra is by definition the tangent space to the group at the identity. The tangent spaces to the group at all other points are identified with the Lie algebra with the help of *right shifts* of the group.

vector fields ξ on M of the form

$$\xi(x) = \frac{d}{dt}\Big|_{t=0} f_t x,$$

where $\{f_t\}$ is a smooth path on $\mathrm{Ham}(M,\Omega)$ with $f_0 = \mathbb{1}$. Every such field is Hamiltonian. Indeed, $\xi = \mathrm{sgrad}\, F_0(x)$, where $F(x,t)$ is the (unique!) normalized Hamiltonian function generating the path, and $F_0(x) = F(x,0)$. Note that $F_0 \in \mathcal{A}$. Vice versa, for every function $F \in \mathcal{A}$ the vector field $\mathrm{sgrad}\, F$ is by definition the derivative at $t = 0$ of the corresponding Hamiltonian flow. We conclude that *the Lie algebra of* $\mathrm{Ham}(M,\Omega)$ *can be identified with* $\mathcal{A}$.

Exercise 1.4.G. Show that with this language the tangent vector to a Hamiltonian path $\{f_t\}$ at $t = s$ is the function $F_s \in \mathcal{A}$. *Hint:* The identification of tangent spaces to the group goes via right shifts (see the footnote). Thus the tangent vector in question is identified with the tangent vector to the path $\{f_t f_s^{-1}\}$ at $t = s$.

The next important notion is the adjoint action of the Lie group on its Lie algebra. Recall that this operation is defined as follows. Pick an element f of the group $\mathrm{Ham}(M,\Omega)$ and an element G from the Lie algebra $\mathcal{A}$. Let $\{g_t\}$, $g_0 = \mathbb{1}$ be a path on the group which is tangent to G. In our situation the tangency condition means of course that the normalized Hamiltonian function of the flow $\{g_t\}$ at time $t = 0$ equals G. By definition,

$$\mathrm{Ad}_f\, G = \frac{d}{dt}\Big|_{t=0} f g_t f^{-1}.$$

Differentiating, we get that the vector field on the right-hand side equals $f_*\,\mathrm{sgrad}\, G$, but this is precisely $\mathrm{sgrad}(G \circ f^{-1})$ in view of Exercise 1.3.B. Returning to our identification we get that

$$\mathrm{Ad}_f\, G = G \circ f^{-1}.$$

Thus *the adjoint action of* $\mathrm{Ham}(M,\Omega)$ *on* $\mathcal{A}$ *is just the usual action of diffeomorphisms on functions.*

Finally let us discuss the Lie bracket on $\mathcal{A}$. Pick two elements $F, G \in \mathcal{A}$, and let $\{f_t\}$, $f_0 = \mathbb{1}$ be a Hamiltonian path tangent to F at 0. The Lie bracket $\{F, G\}$ of F and G is called the Poisson bracket and is defined as follows:

$$\{F, G\} = \frac{d}{dt}\Big|_{t=0} \mathrm{Ad}_{f_t}\, G.$$

Calculating the expression on the right-hand side we get that

$$\{F, G\} = -dG(\mathrm{sgrad}F) = \Omega(\mathrm{sgrad}G, \mathrm{sgrad}F).$$

Let us mention that in terms of vector fields, the Lie bracket coincides with the usual commutator up to the sign. The reader is invited to check that

$$[\operatorname{sgrad} F, \operatorname{sgrad} G] = -\operatorname{sgrad}\{F, G\},$$

where the commutator $[X, Y]$ of two vector fields is defined by $\mathcal{L}_{[X,Y]} = \mathcal{L}_X \mathcal{L}_Y - \mathcal{L}_Y \mathcal{L}_X$.

> **Warning:** Different authors may use different signs in the definitions of the following notions which play an important role in this book: Hamiltonian vector field, Poisson bracket, commutator of vector fields, and curvature of a connection.

Example 1.4.H. Consider the unit sphere S^2 inside the Euclidean space $\mathbb{R}^3$. Let Ω be the induced area form on the sphere. The group $SO(3)$ acts on S^2 by area preserving diffeomorphisms. Since $SO(3)$ is path connected it is contained in $\operatorname{Symp}_0(S^2)$. Applying 1.4.C we get that $\operatorname{Symp}_0(S^2) = \operatorname{Ham}(S^2)$, and thus $SO(3)$ is a subgroup of $\operatorname{Ham}(S^2)$. In particular every element of the Lie algebra $so(3)$ can be uniquely represented as a normalized Hamiltonian function on S^2. We wish to describe this correspondence precisely. Identify $so(3)$ with $\mathbb{R}^3$ as follows. Every vector $a \in \mathbb{R}^3$ is considered as a skew-symmetric transformation $x \to [x, a]$ of the space, where the brackets stand for the standard vector product. Identify the tangent plane to S^2 at a point x with the orthogonal complement to x in the space. By tautological reasons, the Hamiltonian vector field v of the flow $x \to \exp(ta)x$ on the sphere is given by $v(x) = [x, a]$. *We claim that the corresponding normalized Hamiltonian is the height function* $F(x) = (a, x)$. First of all, the reflection over the orthogonal complement to a takes F to $-F$, thus F has zero mean. Further, note that $\Omega(\xi, \eta) = (\eta, [x, \xi])$ for $\xi, \eta \in T_x S^2$. Denote by a' the orthogonal projection of a to $T_x S^2$. Thus $\Omega(\xi, v(x)) = ([x, a'], [x, \xi])$ for every $\xi \in T_x S^2$. Since the vector product with x is an orthogonal transformation of $T_x S^2$, the last expression equals (a', ξ). But this is precisely $dF(\xi)$. The claim follows.

1.5 Algebraic properties of $\operatorname{Ham}(M, \Omega)$

Algebraic properties of the group of Hamiltonian diffeomorphisms were studied by A. Banyaga in [B1], [B2]. In particular, he proved the following striking result. Recall, that a group D is called *simple* if every normal subgroup is trivial, that is equals either $\{\mathbb{1}\}$ or D itself.

Theorem 1.5.A. *Let (M, Ω) be a closed symplectic manifold. Then the group* $\operatorname{Ham}(M, \Omega)$ *is simple.*

There exists a version of this statement for open M as well.

Note that an abelian group is simple if and only if every element generates the whole group (and so it must be a finite cyclic group whose order is a prime number). Thus intuitively speaking general simple groups are far from being abelian. Below we present an elementary statement which clarifies this principle for the group of Hamiltonian diffeomorphisms. We will use it in the next chapter.

Proposition 1.5.B. *Let* (M, Ω) *be a symplectic manifold and let* $U \subset M$ *be a non-empty open subset. There exist* $f, g \in$ Ham(M, Ω) *such that* supp(f), supp$(g) \subset U$ *and* $fg \neq gf$.

The proof is based on the following fact.

Proposition 1.5.C. *Let* $\{f_t\}$ *and* $\{g_t\}$ *be the Hamiltonian flows generated by time independent normalized Hamiltonian functions* F *and* G *respectively. If* $f_t g_t = g_t f_t$ *for all* t *then* $\{F, G\} = 0$.

Proof. According to 1.4.D the corresponding Hamiltonians for $f_t g_t$ and $g_t f_t$ are respectively

$$F(x) + G(f_t^{-1}(x)) \quad \text{and} \quad G(x) + F(g_t^{-1}(x)).$$

But since the flows are the same for both we have that

$$F(x) + G(f_t^{-1}(x)) = G(x) + F(g_t^{-1}(x))$$

for all t. Differentiating with respect to t we get

$$dG(-\text{sgrad}F) = dF(-\text{sgrad}G)$$

which yields $\{F, G\} = \{G, F\}$. Hence, from the anti-commutativity of the Lie bracket, $\{F, G\} = 0$. $\qquad \square$

Proof of 1.5.B. Choose a point $x \in U$ and tangent vectors $\xi, \eta \in T_x U$ such that $\Omega(\xi, \eta) \neq 0$. Now choose germs of functions F and G (see exercise below) satisfying sgrad$F(x) = \xi$, sgrad$G(x) = \eta$. Extend these functions by 0 outside U. If M is open we are done. If M is closed, add a constant to guarantee that F and G have zero mean. So now the functions F, G belong to $\mathcal{A}$. Further, they are constant outside U, and so the corresponding Hamiltonian diffeomorphisms f_t and g_t are supported in U. Since $\{F, G\} \neq 0$ we see that for some t, the diffeomorphisms f_t and g_t do not commute. $\qquad \square$

Exercise. Use local canonical coordinates near x to show that F and G in the proof of the corollary do exist.

We complete this section with the following result due to A. Banyaga (along the lines of [B2]).

Theorem 1.5.D. *Let (M_1, Ω_1) and (M_2, Ω_2) be two closed symplectic manifolds such that their groups of Hamiltonian diffeomorphisms are isomorphic. Then the manifolds are* conformally symplectomorphic: *there exists a diffeomorphism $f : M_1 \to M_2$ and a number $c \neq 0$ such that $f^* \Omega_2 = c\Omega_1$.*

In other words, the algebraic structure of the group of Hamiltonian diffeomorphisms determines the symplectic manifold up to a factor.

Chapter 2

Introducing the Geometry

In the present chapter we discuss biinvariant Finsler metrics on the group of Hamiltonian diffeomorphisms and define Hofer's geometry.

2.1　A variational problem

What is the minimal amount of energy needed in order to generate a given Hamiltonian diffeomorphism ϕ? This natural question can be formalized as follows. Consider all possible Hamiltonian flows $\{f_t\}$, $t \in [0;1]$ such that $f_0 = \mathbb{1}$ and $f_1 = \phi$. For each flow take its unique normalized Hamiltonian function $F_t(x)$ and "measure its magnitude". Then minimize the result of the measurement over all such flows. It remains to explain what we mean by the "magnitude". Recall that for every t the function F_t is an element of the Lie algebra $\mathcal{A}$. Choose any norm $||\ ||$ on $\mathcal{A}$ in coordinate-free way, that is we require that

$$||H \circ \psi^{-1}|| = ||H|| \text{ for all } H \in \mathcal{A}, \ \psi \in \mathrm{Ham}(M, \Omega). \tag{2.1.A}$$

Now define the magnitude as $\int_0^1 ||F_t|| dt$. Bringing together all ingredients of this procedure we end up with the variational problem

$$\inf \int_0^1 ||F_t|| dt, \tag{2.1.B}$$

where ϕ is given, and the infimum is taken over all flows $\{f_t\}$ as above.

2.2 Biinvariant geometries on $\mathrm{Ham}(M, \Omega)$

It turns out that the variational problem above can be reformulated in purely geometrical terms. To make this transparent I wish to recall the notion of *a Finsler structure* on a manifold. We say that a manifold Z is endowed with a Finsler structure if its tangent spaces $T_z Z$ are equipped with a norm which depends smoothly on the points $z \in Z$. Of course, Riemannian structures form a particular case of this notion. In general however the norms above may not come from a scalar product. Given a Finsler structure, one defines *the length of a curve* exactly as in the Riemannian case by

$$\mathrm{length}\{z(t)\}_{t \in [a;b]} = \int_a^b \mathrm{norm}(\dot{z}(t))dt.$$

Further, one introduces *the distance* between two points z and z' in Z as the infimum of lengths of all curves joining z and z'.

Let us return to the situation described in the previous section. Since all tangent spaces to the group $\mathrm{Ham}(M, \Omega)$ are identified with $\mathcal{A}$ (see section 1.4 above), every choice of the norm $\| \ \|$ on $\mathcal{A}$ leads to a Finsler structure on the group. Thus one can define the length of a Hamiltonian path, and the distance between two Hamiltonian diffeomorphisms. In particular, the length of a Hamiltonian path $\{f_t\}$, $t \in [a;b]$ with the normalized Hamiltonian F is given by

$$\mathrm{length}\{f_t\} = \int_a^b \|F_t\|dt.$$

The distance between two Hamiltonian diffeomorphisms ϕ and ψ is defined by

$$\rho(\phi, \psi) = \inf \mathrm{length}\{f_t\},$$

where the infimum is taken over all Hamiltonian paths $\{f_t\}$, $t \in [a;b]$ with $f_a = \phi$ and $f_b = \psi$. Of course the length of a path does not depend on the parametrization, thus in the definition of the distance above one can take $a = 0$ and $b = 1$. With this language, the solution of the variational problem 2.1.B above is nothing else but the distance $\rho(\mathbb{1}, \phi)$!

The following properties of ρ are easily verified and are left as an exercise:

- $\rho(\phi, \psi) = \rho(\psi, \phi)$;
- the triangle inequality : $\rho(\phi, \psi) + \rho(\psi, \theta) \geq \rho(\phi, \theta)$;
- $\rho(\phi, \psi) \geq 0$.

Recall now Condition 2.1.A imposed on the norm $\| \ \|$ in section 2.1 above. In geometric language this condition means that the norm is invariant under the adjoint action of the group on the Lie algebra (see 1.4 above). In what follows we deal with such norms only.

Exercise. Show that 2.1.A yields that the function ρ is *biinvariant:*[1]

$$\rho(\phi, \psi) = \rho(\phi\theta, \psi\theta) = \rho(\theta\phi, \theta\psi)$$

for all $\phi, \psi, \theta \in \mathrm{Ham}(M, \Omega)$.

It would be more honest to call the function ρ a *pseudo-distance.* Indeed, as we have seen above it satisfies all the axioms of a distance function on a metric space except possibly *the non-degeneracy,* that is

$$\rho(\phi, \psi) > 0 \text{ for } \phi \neq \psi. \tag{2.2.A}$$

It is rather non-trivial to check the non-degeneracy even in finite dimensional geometry. The argument in this case uses local compactness of manifolds. In our situation the group $\mathrm{Ham}(M, \Omega)$ is infinite dimensional and has no compactness properties. Thus a priori there is no reason for 2.2.A to be true. It turns out that the non-degeneracy of ρ is very sensitive to the choice of the norm $\| \ \|$.

2.3   The choice of the norm: L_p vs. L_∞

Among the norms on $\mathcal{A}$ satisfying the invariancy assumption 2.1.A there is a very natural class which includes L_p norms, $p = 1, 2, 3, \ldots$

$$\|H\|_p = \left(\int_M |H|^p \mathrm{Vol} \right)^{\frac{1}{p}},$$

and the L_∞-norm

$$\|H\|_\infty = \max H - \min H.$$

Denote by ρ_p and ρ_∞ the corresponding pseudo-distances.

Theorem 2.3.A. *The pseudo-distance ρ_p is degenerate for all finite $p = 1, 2, \ldots$. Moreover, if the manifold is closed then all such ρ_p vanish identically.*

This result was established in [EP]. The proof is presented below in this chapter (see also books [HZ], [MS], [AK]). The next theorem shows a striking contrast between the L_p and the L_∞ cases.

---

[1]Without assumption 2.1.A we get only right-invariant ρ's, that is $\rho(\phi, \psi) = \rho(\phi\theta, \psi\theta)$. Such metrics play an important role in hydrodynamics, see [AK].

Theorem 2.3.B. *The pseudo-distance ρ_∞ is non-degenerate.*

This theorem[2] was discovered and proved by Hofer in [H1] for the case $M = \mathbb{R}^{2n}$ with the use of infinite dimensional variational methods. In [V1] Viterbo derived it for the case $M = \mathbb{R}^{2n}$ from his theory of generating functions. Both Hofer and Viterbo obtained their results as an answer to a stimulating question posed by Eliashberg in a private discussion. In [P1] the statement was extended to a wide class of symplectic manifolds with "nice" behaviour at infinity, and in particular to all closed symplectic manifolds such that the cohomology class of the symplectic form is rational. The approach of [P1] is based on Gromov's theory of pseudo-holomorphic curves. Finally [LM1] Lalonde and McDuff proved 2.3.B in full generality using Gromov's theory. At present some other proofs of various particular cases of this theorem are available, see for instance [Ch],[O3],[Sch3]. Hofer's original proof is presented in great details in the book [HZ]. The argument due to Lalonde and McDuff is outlined in the book [MS] and in the survey [L]. Below we give a different proof for the case of $M = \mathbb{R}^{2n}$ which follows [P1]. All known proofs are based on "hard" methods.[3]

2.4 The concept of displacement energy

Which invariant norms $\|\ \|$ on $\mathcal{A}$ (that is norms satisfying 2.1.A) lead to non-degenerate distance functions ρ? Below we describe a very useful reformulation of this question which will eventually enable us to prove Theorem 2.3.A. It is based on the beautiful concept of displacement energy which was introduced by Hofer in [H1]. Let ρ be a biinvariant pseudo-distance on $\mathrm{Ham}(M,\Omega)$, and let A be a bounded subset of M.

Definition. The *displacement energy* of A is given by

$$e(A) = \inf\{\rho(\mathbb{1}, f) \mid f \in \mathrm{Ham}(M,\Omega),\ f(A) \cap A = \emptyset\}.$$

The set of such f may be empty. In what follows we use the convention that the infimum of the empty set equals $+\infty$. If $e(A) \neq 0$ we will say that A has positive displacement energy.

  Let us mention two obvious but important properties of $e(A)$. First of all, e is a monotone function of subsets: if $A \subset B$ then $e(A) \leq e(B)$. Second, e

---

[2]The historical digression below reflects my own understanding of the situation. I admit that other participants of this development may see it differently.

[3]Moreover, to my taste all the proofs are far from being transparent. The argument presented in Chapter 3 below is not an exception. I strongly believe that in the future one will find *the explanation* of this fundamental result.

is an invariant, that is $e(A) = e(f(A))$ for every Hamiltonian diffeomorphism f of M. We leave the proofs to the reader.

Example. Consider $(\mathbb{R}^2, \omega)$ and take an open square A whose edges have length u and are parallel to the coordinate axes. Let us estimate the displacement energy of A with respect to the distance ρ_∞. Consider the Hamiltonian function $H(p, q) = up$. The corresponding Hamiltonian system is

$$\begin{cases} \dot{q} = u \\ \dot{p} = 0 \end{cases}$$

Therefore its time-1-map h sends (p, q) to $(p, q+u)$. Note that all the motion of the square takes place in the rectangle $K = \text{Closure}(A \cup h(A))$. Consider a cut off F of the Hamiltonian H outside a small neighbourhood of K.[4] Note that F (in contrast to H) is a normalized Hamiltonian function. Since $F = H$ on K, the time one map f of the Hamiltonian flow generated by F still displaces the square A. We can always perform the cutting off in such a way that the L_∞-norm of F is arbitrary close to the oscillation of H on K. This oscillation equals

$$\max_K H - \min_K H = u^2 - 0 = u^2.$$

Thus we get that

$$e(A) \leq u^2 = \text{Area}(A).$$

Note that the square is symplectomorphic to the disc of the same area in $\mathbb{R}^2$ (this is not true in higher dimensions!). Thus we proved that $e(B^2(r)) \leq \pi r^2$. A deep result due to Hofer ([H1]) states that actually there is equality in all dimensions, in other words $e(B^{2n}(r)) = \pi r^2$. See [LM1] for a generalization to arbitrary symplectic manifolds. We will prove a lower bound for $e(B^{2n}(r))$ in the next chapter.

In general, if the displacement energy of all non-empty open subsets with respect to some biinvariant pseudo-metric ρ is positive then ρ is non-degenerate. Indeed, each $f \in \text{Ham}(M, \Omega)$ such that $f \neq \mathbb{1}$ must displace some small ball $A \subset M$. Thus we get that $\rho(\mathbb{1}, f) \geq e(A) > 0$. In fact, the converse is also true.

Theorem 2.4.A. *([EP]) If ρ is non-degenerate then $e(A) > 0$ for every non-empty open subset A.*

In order to prove this theorem we need an auxiliary lemma.

---

[4]Let Y be a closed subset of a manifold Z, and let H be a smooth function defined in a neighbourhood V of Y. By a cut off F of H outside a small neigbourhood of Y we mean the following. Choose a neighbourhood W of Y whose closure is contained in V. Take a smooth function $a : Z \to [0; 1]$ which equals 1 on W and vanishes outside V. Define now F by aH on V and by 0 outside V.

Lemma 2.4.B. *Let $A \subset M$ be a non-empty open subset. For all $\phi, \psi \in$ Ham(M, Ω) with $\mathrm{supp}(\phi) \subset A$ and $\mathrm{supp}(\psi) \subset A$, we have $e(A) \geq \frac{1}{4}\rho(\mathbb{1}, [\phi, \psi])$.*

Here $[\phi, \psi]$ stands for the commutator $\psi^{-1}\phi^{-1}\psi\phi$. The theorem immediately follows from the lemma in view of Proposition 1.5.B of the previous chapter.

Proof of 2.4.A. By 1.5.B there exist ϕ, ψ supported in A with $[\phi, \psi] \neq \mathbb{1}$. Since ρ is non-degenerate this implies that $e(A) \geq \frac{1}{4}\rho(\mathbb{1}, [\phi, \psi]) > 0$.   □

Proof of 2.4.B. Assume that there exists $h \in$ Ham(M, Ω) such that $h(A) \cap A = \emptyset$ (if such an h does not exist we are done because $e(A) = +\infty$). Set $\theta = \phi h^{-1}\phi^{-1}h = [h, \phi^{-1}]$. If $x \in A$ then $h(x) \notin A$. Since $\phi = \mathbb{1}$ outside A we get $\phi^{-1}h(x) = h(x)$. So we see that $h^{-1}\phi^{-1}h(x) = x$ and therefore $\theta|_A = \phi|_A$. Now $\mathrm{supp}(\psi) \subset A$ and therefore $\phi^{-1}\psi\phi = \theta^{-1}\psi\theta$ which implies that $[\phi, \psi] = [\theta, \psi]$. Note that

$$\begin{aligned}
\rho(\mathbb{1}, [\theta, \psi]) &= \rho(\mathbb{1}, \psi^{-1}\theta^{-1}\psi\theta) \\
&= \rho(\theta^{-1}, \psi^{-1}\theta^{-1}\psi) \\
&\leq \rho(\mathbb{1}, \theta^{-1}) + \rho(\mathbb{1}, \psi^{-1}\theta^{-1}\psi) \\
&= 2\rho(\mathbb{1}, \theta).
\end{aligned}$$

Here we have used the bi-invariance of ρ and the triangle inequality. Analogously,

$$\rho(\mathbb{1}, \theta) = \rho(\mathbb{1}, [h, \phi^{-1}]) \leq 2\rho(\mathbb{1}, h).$$

Putting these two inequalities together we get

$$\rho(\mathbb{1}, [\phi, \psi]) = \rho(\mathbb{1}, [\theta, \psi]) \leq 2\rho(\mathbb{1}, \theta) \leq 4\rho(\mathbb{1}, h).$$

Since this holds for all $h \in$ Ham(M, Ω) with $h(A) \cap A = \emptyset$ we obtain, by taking the infimum, that $4e(A) \geq \rho(\mathbb{1}, [\phi, \psi])$.   □

Recall that Theorem 2.3.A states that for $p < \infty$ the L_p-norm leads to a degenerate pseudo-distance, which in fact vanishes on closed manifolds. Now we will prove this statement and see why the argument breaks down in the L_∞-case.

Proof of 2.3.A. We will show that the displacement energy of a "small" ball vanishes. Then the degeneracy of ρ_p follows from 2.4.A. Let U be an open subset of M endowed with canonical coordinates (x, y). In these coordinates the symplectic form Ω is given by $\sum dx_i \wedge dy_i$. Assume without loss of generality that U contains a ball $\sum(x_j^2 + y_j^2) < 10$. Let $A \subset U$ be the ball with the same center of the radius 0.1. Consider a (partially defined) flow $h_t, t \in [0; 1]$ on U which is simply a shift by t along the y_1 coordinate. Such a shift is generated by a (non-normalized!) Hamiltonian function $H(x,y) = x_1$ on U. Clearly, $h_1(A) \cap A = \emptyset$. Denote by S_t the sphere $h_t(\partial A)$. Consider a new (time-dependent) normalized Hamiltonian function $G_t = F_t + c_t$ where F_t is obtained from H by a cut off outside a small neighbourhood of S_t, and c_t is a (time dependent) constant. Here of course $c_t = 0$ if the manifold M is open, and $c_t = -\mathrm{Vol}(M)^{-1} \int_M F_t \mathrm{Vol}$ if M is closed. Since for every t the function G_t coincides with H near S_t up to an additive constant we conclude that $\mathrm{sgrad}\, G_t = \mathrm{sgrad}\, H$ near S_t. Hence the Hamiltonian flow $\{g_t\}$ of G satisfies $g_t(\partial A) = h_t(\partial A)$ and so $g_1(\partial A) \cap \partial A = \emptyset$. But this obviously implies that $g_1(A) \cap A = \emptyset$. Note now that using cutting off outside very small neighbourhoods of S_t we can achieve that the L_p-norm of every function G_t is arbitrary small (in contrast to the L_∞-norm!). Therefore the L_p- displacement energy of A vanishes. This completes the proof of the degeneracy of ρ_p.

Let us turn now to the second statement of the theorem, where we assume that the manifold M is closed. Set

$$\mathcal{G} = \{g \in \mathrm{Ham}(M, \Omega) \mid \rho(\mathbb{1}, g) = 0\}.$$

Take $f, g \in \mathcal{G}$. Then of course $g^{-1} \in \mathcal{G}$. Further the triangle inequality yields that

$$\rho(\mathbb{1}, fg) = \rho(f^{-1}, g) \leq \rho(\mathbb{1}, f) + \rho(\mathbb{1}, g) = 0$$

so $\mathcal{G}$ is a subgroup of $\mathrm{Ham}(M, \Omega)$. By the bi-invariance we know that if $f \in \mathcal{G}$ and $h \in \mathrm{Ham}(M, \Omega)$ then $hfh^{-1} \in \mathcal{G}$, and so $\mathcal{G}$ is a normal subgroup. In view of Banyaga's theorem 1.5.A above the group $\mathrm{Ham}(M, \Omega)$ is simple. Therefore either $\mathcal{G} = \{\mathbb{1}\}$ or $\mathcal{G} = \mathrm{Ham}(M, \Omega)$. We already proved that ρ_p is degenerate, hence $\mathcal{G} \neq \{\mathbb{1}\}$. Thus $\mathcal{G}$ coincides with the whole group $\mathrm{Ham}(M, \Omega)$. We conclude that ρ_p vanishes identically. $\qquad\square$

Exercise. Prove that the displacement energy of $S^{2n-2} \subset \mathbb{R}^{2n-1} \subset \mathbb{R}^{2n}$ with respect to ρ_∞ vanishes. On the other hand, we will see in the next chapter that there exist half-dimensional submanifolds of $\mathbb{R}^{2n}$ which have positive displacement energy (cf. 1.1.C above).

Open problem. Which invariant norms on $\mathcal{A}$ give rise to non-degenerate distance functions ρ? Is it true that such norms are always bounded below by $\mathrm{const}\| \; \|_\infty$? A difficulty here is that no classification of $\mathrm{Ham}(M, \Omega)$-invariant norms is known. A potential approach to this problem would be to investigate cut-offs. If cutting off decreases the norm arbitrarily then our argument above shows that ρ is degenerate.

Open problem. [EP] It is quite natural to consider separately the positive and the negative parts of the metric ρ_∞. More explicitly, set

$$\rho_+(\mathbb{1}, f) = \inf \int_0^1 \max F_t \, dt,$$

and

$$\rho_-(\mathbb{1}, f) = \inf \int_0^1 - \min F_t dt.$$

Then clearly

$$\rho(\mathbb{1}, f) \geq \rho_+(\mathbb{1}, f) + \rho_-(\mathbb{1}, f).$$

However in all examples known to me equality holds! It would be interesting to prove a general statement, or to find a counterexample. Note that it follows from [V1] that on $\mathrm{Ham}(\mathbb{R}^{2n})$ the sum $\rho_+ + \rho_-$ defines a biinvariant metric. As far as I know, no analogue of this statement was proved for general symplectic manifolds.

Convention. Unless otherwise stated in what follows we use notation $\| \ \|$ for the L_∞-norm on $\mathcal{A}(M)$. We write ρ for the metric ρ_∞ and call it *Hofer's metric*. The quantity $\rho(\mathbb{1}, f)$ is called *Hofer's norm* of f. We write length$\{f_t\}$ for the length of a Hamiltonian path $\{f_t\}$ with respect to the L_∞-norm (see 2.2 above).

Chapter 3

Lagrangian Submanifolds

The purpose of the next two chapters is to prove that Hofer's metric on $\mathbb{R}^{2n}$ is non-degenerate. We use an approach of [P1]. For that purpose we introduce Lagrangian submanifolds of symplectic manifolds. Lagrangian submanifolds play a fundamental role in symplectic topology as well as in its applications to mechanics and calculus of variations. They will appear on many occasions throughout this book.

3.1 Definitions and examples

Definition. Let (M^{2n}, Ω) be a symplectic manifold and let $L \subset M$ be a submanifold. We call L *Lagrangian* if $\dim L = \frac{1}{2}\dim M = n$ and $\Omega|_{TL} \equiv 0$. An embedding (or immersion) $f : L^n \to M^{2n}$ is called Lagrangian if $f^*\Omega \equiv 0$.

Let us list some important examples of Lagrangian submanifolds.

3.1.A. Curves on surfaces
Let (M^2, Ω) be an oriented surface with an area form. Then every curve is Lagrangian (since the tangent space to a curve is one dimensional and Ω vanishes when it is evaluated on two vectors which are proportional).

3.1.B. The split torus (cf. 1.1.C above).
The torus $S^1 \times \cdots \times S^1 \subset \mathbb{R}^2 \times \cdots \times \mathbb{R}^2 = \mathbb{R}^{2n}$ is Lagrangian (the 2-form Ω on $\mathbb{R}^{2n}$ splits).

3.1.C. Graphs of closed 1-forms in cotangent bundles
This example plays an important role in classical mechanics. Let N^n be any manifold. Consider the cotangent bundle $M = T^*N$ of N with the natural projection $\pi : T^*N \to N, (p, q) \mapsto q$. Define the following 1-form λ on M which is called the Liouville form. For $q \in N, (p, q) \in T^*N$ and $\xi \in T_{(p,q)}T^*N$

we set $\lambda(\xi) = \langle p, \pi_*\xi \rangle$ where $\langle \, , \, \rangle$ is the natural pairing between T^*N and TN. We claim that $\Omega = d\lambda$ is a symplectic form on T^*M. We use local coordinates $(p_1, \ldots, p_n, q_1, \ldots, q_n)$ on T^*N and write $\xi = (\dot{p}_1, \ldots, \dot{p}_n, \dot{q}_1, \ldots, \dot{q}_n)$, so $\pi_*\xi = (\dot{q}_1, \ldots, \dot{q}_n)$. With this notation the pairing reads $\langle p, \pi_*\xi \rangle = \sum p_i \dot{q}_i$ which implies that $\lambda(\xi) = \sum p_i dq_i$. Therefore $\Omega = d\lambda = \sum dp_i \wedge dq_i$, and we recognize the standard symplectic form on $\mathbb{R}^{2n}$. The claim follows.

Exercise. Let α be a 1-form on N. Show that graph(α) is a Lagrangian submanifold of T^*N if and only if α is closed.

3.1.D. Symplectomorphisms as Lagrangian submanifolds
Let $f : (M, \Omega) \to (M, \Omega)$ be a diffeomorphism. Consider a new symplectic manifold $(M \times M, \Omega \oplus -\Omega)$. We leave it as an exercise to show that graph $(f) \subset (M \times M, \Omega \oplus -\Omega)$ is Lagrangian if and only if f is a symplectomorphism.

3.1.E. Lagrangian suspension
Let $L \subset (M, \Omega)$ be a Lagrangian submanifold. Consider a loop of Hamiltonian diffeomorphisms $\{h_t\}$, $t \in S^1$, $h_0 = h_1 = \mathbb{1}$ generated by a 1-periodic Hamiltonian function $H(x, t)$.

Proposition. *Let $M \times T^*S^1$ be a symplectic manifold with symplectic form $\sigma = \Omega + dr \wedge dt$ where M is as above and (r, t) are the coordinates on $T^*S^1 = \mathbb{R} \times S^1$. Then*

$$\phi : L \times S^1 \to M \times T^*S^1, (x, t) \mapsto (h_t(x), -H(h_t(x), t), t)$$

is a Lagrangian embedding.

Proof. It suffices to prove that $\phi^*\sigma$ vanishes on pairs (ξ, ξ') and on pairs $(\xi, \frac{\partial}{\partial t})$ for $\xi, \xi' \in TL$ and $\frac{\partial}{\partial t} \in TS^1$. One computes that

$$\phi_*\xi = h_{t*}\xi - \langle dH_t, h_{t*}\xi \rangle \frac{\partial}{\partial r} \, ,$$

$$\phi_*\xi' = h_{t*}\xi' - \langle dH_t, h_{t*}\xi' \rangle \frac{\partial}{\partial r} \, ,$$

so $\phi^*\sigma(\xi, \xi') = \Omega(h_{t*}\xi, h_{t*}\xi') = \Omega(\xi, \xi') = 0$ since L is Lagrangian. Furthermore

$$\phi_*\frac{\partial}{\partial t} = \text{sgrad } H_t - \left(\langle dH_t, \text{sgrad } H_t \rangle + \frac{\partial H}{\partial t} \right) \frac{\partial}{\partial r} + \frac{\partial}{\partial t}$$

$$= \text{sgrad } H_t - \frac{\partial H}{\partial t} \frac{\partial}{\partial r} + \frac{\partial}{\partial t} \, ,$$

so that we get

$$\phi^*\Omega(\xi, \frac{\partial}{\partial t}) = \Omega(h_{t*}\xi, \operatorname{sgrad} H_t) + dr \wedge dt(-\langle dH_t, h_{t*}\xi\rangle \frac{\partial}{\partial r}, \frac{\partial}{\partial t})$$
$$= \Omega(h_{t*}\xi, \operatorname{sgrad} H_t) - \langle dH_t, h_{t*}\xi\rangle$$
$$= dH_t(h_{t*}\xi) - \langle dH_t, h_{t*}\xi\rangle$$
$$= 0. \qquad \qquad \square$$

3.2 The Liouville class of Lagrangian submanifolds in $\mathbb{R}^{2n}$

Let $L \subset (\mathbb{R}^{2n}, dp \wedge dq)$ be a Lagrangian submanifold. Consider the restriction $\lambda|_{TL}$ of the Liouville form

$$\lambda = p_1 dq_1 + \cdots + p_n dq_n$$

to L. Clearly $d(\lambda|_{TL}) = \Omega|_{TL} = 0$. The cohomology class $\lambda_L \in H^1(L, \mathbb{R})$ of this closed 1-form is called *the Liouville class* of the Lagrangian submanifold L. Similary, for a Lagrangian embedding or immersion $\phi : L \to \mathbb{R}^{2n}$ we define the Liouville class as $[\phi^*\lambda]$. Geometrically, the Liouville class of a Lagrangian submanifold can be interpreted as follows. Let $a \in H_1(L)$ be a 1-cycle. Then we can find a 2-chain Σ in $\mathbb{R}^{2n}$ with $\partial\Sigma = a$. Now

$$(\lambda_L, a) = \int_a \lambda_L = \int_\Sigma \Omega$$

and it is easy to see that this is independent of the choice of Σ. In view of this formula the value (λ_L, a) is sometimes called *the symplectic area* of a. The last construction generalizes to an arbitrary Lagrangian manifold L of a symplectic manifold M and defines a natural homomorphism $H_2(M, L; \mathbb{Z}) \to \mathbb{R}$. An important property of λ_L is that it is invariant under symplectomorphisms of $\mathbb{R}^{2n}$ i.e. $f^*\lambda_{f(L)} = \lambda_L$.

Theorem 3.2.A. *([G1]) Assume that $L \subset \mathbb{R}^{2n}$ is a closed Lagrangian submanifold. Then $\lambda_L \neq 0$.*

Let us emphasize that here L is embedded. For Lagrangian immersions this statement is in general not true. In the case $n = 1$ this can be seen as follows. It is clear that any closed embedded curve bounds a subset of positive area, but for example the immersed figure eight curve can bound a set with zero area. This phenomenon reflects "rigidity of Lagrangian embeddings".

Definition. A closed Lagrangian submanifold $L \subset (\mathbb{R}^{2n}, \omega)$ is called *rational* if $\lambda_L(H_1(L; \mathbb{Z})) \subset \mathbb{R}$ is a discrete subgroup.

We will denote its positive generator by $\gamma(L)$.

Example. The split torus $L = S^1(r) \times \cdots \times S^1(r) \subset \mathbb{R}^{2n}$ is rational. Since each $S^1(r)$ has symplectic area πr^2 we have $\gamma(L) = \pi r^2$. However, the torus $S^1(1) \times S^1(\sqrt[3]{2}) \subset \mathbb{R}^4$ is not rational. The symplectic areas of the two circles are respectively π and $\sqrt[3]{4}\pi$ and they generate a dense subgroup of $\mathbb{R}$.

Theorem 3.2.B. *([S1]) Let $L \subset B^2(r) \times \mathbb{R}^{2n-2}$ be a closed rational Lagrangian submanifold. Then $\gamma(L) \leq \pi r^2$.*

Note that the assumption that L is embedded is necessary. Figure 1 shows an immersed Lagrangian submanifold of arbitrary symplectic area.

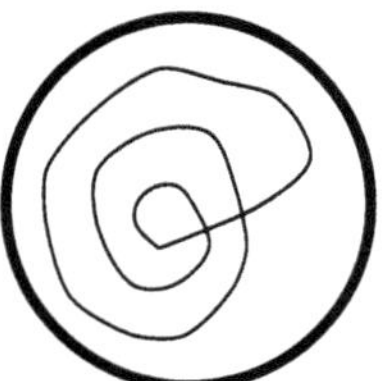

Figure 1

Our next result gives a lower bound for the displacement energy $e(L)$ of a rational Lagrangian submanifold L with respect to Hofer's metric.

Theorem 3.2.C. *Let $L \subset \mathbb{R}^{2n}$ be a closed rational Lagrangian submanifold. Then $e(L) \geq \frac{1}{2}\gamma(L)$.*

Theorems 3.2.A and 3.2.B are proved in the next chapter. Theorem 3.2.C follows from 3.2.B (see section 3.3 below). Let us derive some consequences of these results.

3.2.D. Non-degeneracy of Hofer's metric. Theorem 3.2.C implies that Hofer's metric on $\mathrm{Ham}(\mathbb{R}^{2n}, \omega)$ is non-degenerate. Indeed, each ball $B^{2n}(r) = \{p_1^2 + \cdots + p_n^2 + q_1^2 + \cdots + q_n^2 \leq r^2\}$ contains a rational split torus

$$S^1(\frac{r}{\sqrt{n}}) \times \cdots \times S^1(\frac{r}{\sqrt{n}}) = \left\{ p_1^2 + q_1^2 = \frac{r^2}{n}, \ldots, p_n^2 + q_n^2 = \frac{r^2}{n} \right\}.$$

Thus $e(B^{2n}(r)) \geq \frac{\pi r^2}{2n} > 0$, and as it was explained in 2.4 above we get the desired non-degeneracy of ρ. This estimate is not sharp. Hofer proved [H1] that $e(B^{2n}(r)) = \pi r^2$.

3.2.E. The non-squeezing property. Note that $\gamma(L)$ is a symplectic invariant i.e. for a symplectomorphism $f : \mathbb{R}^{2n} \to \mathbb{R}^{2n}$ we have $\gamma(f(L)) = \gamma(L)$. Thus Theorem 3.2.B implies the non-squeezing Theorem 1.1.C. Recall that it states that a split torus with large $\gamma(L) = \pi R^2$ cannot be moved by a Hamiltonian diffeomorphism into $B^2(r) \times \mathbb{R}^{2n-2}$ for $r < R$.

3.2.F. The cylindrical symplectic capacity. Let $A \subset \mathbb{R}^{2n}$ be a bounded subset and set

$$c(A) = \inf\{\pi r^2 | \exists\, g : \mathbb{R}^{2n} \to \mathbb{R}^{2n} \text{with } g(A) \subset B^2(r) \times \mathbb{R}^{2n-2}\} \,,$$

where g is a symplectomorphism. This function defined on subsets of $\mathbb{R}^{2n}$ is called the cylindrical symplectic capacity. In this language Theorem 3.2.B reads that for a closed rational Lagrangian submanifold $c(L) \geq \gamma(L)$. This capacity is a symplectic invariant and satisfies the following monotonicity property. For $A \subset B$, $c(A) \leq c(B)$ (compare to a similar monotonicity property of the displacement energy, see 2.4 above).

3.2.G. Some generalizations. Let (M, Ω) be a symplectic manifold. When M is open we assume that it has a "nice" behaviour at infinity (this class includes for instance any cotangent bundle endowed with the standard symplectic structure, as well as a product of a cotangent bundle with any closed symplectic manifold). Take a Lagrangian submanifold $L \subset M$, and consider a homomorphism $\lambda_L : \pi_2(M, L) \to \mathbb{R}$ which takes any disc Σ in M whose boundary lies on L to its symplectic area $\int_\Sigma \Omega$. Exactly as in the case $M = \mathbb{R}^{2n}$ we say that L is rational if the image of λ_L is discrete, and for a rational L define $\gamma(L)$ as the positive generator of the image of λ_L. When $\lambda_L = 0$ put $\gamma(L) = +\infty$. Our proof of Theorem 3.2.C extends without essential changes to this more general setting (see [P1]), namely one gets that $e(L) \geq \frac{1}{2}\gamma(L)$. As a corollary,[1] we obtain the following important fact proved by Gromov in [G1]: $e(L) = +\infty$ when $\lambda_L = 0$. This can be interpreted as a Lagrangian intersection property: if $\lambda_L = 0$ then $\phi(L)$ intersects L for every Hamiltonian diffeomorphism ϕ. We discuss an application of this result to Hofer's geometry in Chapter 6 below.

Let us mention also that these estimates were significantly improved in [Ch] with the use of Floer homology (see also [O3]). In particular it was shown that every (not necessarily rational) closed Lagrangian submanifold $L \subset M$ has positive displacement energy.

3.2.H. An isoperimetric inequality. We conclude this section with the following beautiful result due to Viterbo [V2]. Let $L \subset \mathbb{R}^{2n}$ be a closed Lagrangian submanifold. Denote by V the n-dimensional Euclidean volume of L. Then

$$e(L) \leq 2^{n(n-1)/2} n^n V.$$

The precise constant is still to be found.

---

[1] This point was missed in [P1, p. 359].

3.3   Estimating the displacement energy

In this section we deduce Theorem 3.2.C from Theorem 3.2.B by an elementary geometric argument.

Step 1. Let L be a closed rational Lagragian submanifold and let $h_t, t \in [0, 1]$ be a path of Hamiltonian diffeomorphisms such that $h_0 = \mathbb{1}$ and $h_1(L) \cap L = \emptyset$. Fix $\varepsilon > 0$. Without loss of generality one can assume that $h_t = \mathbb{1}$ for $t \in [0, \varepsilon]$ and $h_t = h_1$ for $t \in [1 - \varepsilon, 1]$. Indeed, one can achieve this by a suitable reparametrization of the flow which preserves its length (use Exercise 1.4.A above). Let $H(x, t)$ be the corresponding Hamiltonian function. Set

$$l = \text{length } \{h_t\} = \int_0^1 \max_x H_t - \min_x H_t dt.$$

We have to prove that $l \geq \frac{1}{2}\gamma(L)$. The master plan is to encode the motion of L under the flow as a closed Lagrangian submanifold in $\mathbb{R}^{2n+2}$ and then obtain the result using 3.2.B. We will make use of the Lagrangian suspension construction described before. In order to do this we need a loop of Hamiltonian diffeomorphisms.

Step 2. We create a loop of Hamiltonian diffeomorphisms for $t \in [0, 2]$

$$g_t = \begin{cases} h_t & \text{for } t \in [0, 1] \\ h_{2-t} & \text{for } t \in [1, 2] \end{cases}$$

with corresponding Hamiltonian

$$G(x, t) = \begin{cases} H(x, t) & \text{for } t \in [0, 1] \\ -H(x, 2 - t) & \text{for } t \in [1, 2]. \end{cases}$$

Exercise. Show that $\int_0^2 G(g_t(x), t)dt = 0$ for all x.

Apply now the Lagrangian suspension construction (see 3.1.E above). We get a new Lagrangian submanifold $L' \subset \mathbb{R}^{2n} \times T^*S^1$ as the image of $L \times S^1$ under the mapping

$$(x, t) \mapsto (g_t(x), -G(g_t(x), t), t).$$

Note that we take here $S^1 = \mathbb{R}/2\mathbb{Z}$. Define two functions

$$a_+(t) = -\min_x G(x, t) + \varepsilon \quad \text{and} \quad a_-(t) = -\max_x G(x, t) - \varepsilon.$$

Clearly $L' \subset \mathbb{R}^{2n} \times C \subset \mathbb{R}^{2n} \times T^*S^1$, where C stands for the annulus $\{a_-(t) < r < a_+(t)\}$ (see Figure 2).

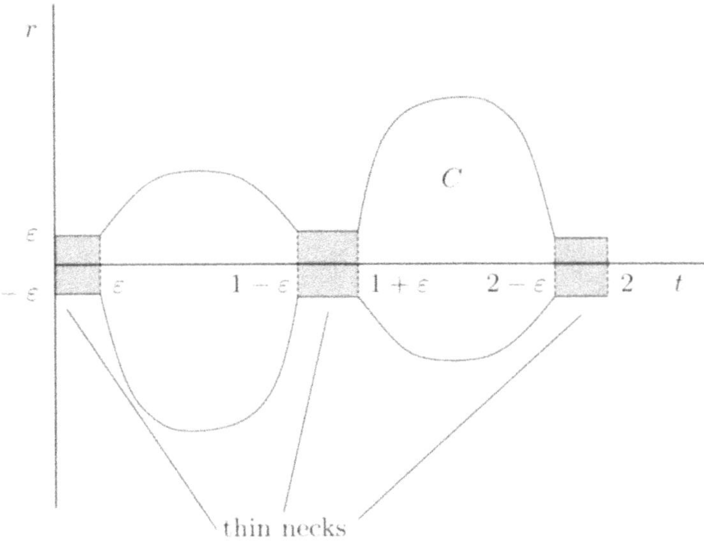

Figure 2

Step 3. We now wish to pass from $\mathbb{R}^{2n} \times T^*S^1$ to $\mathbb{R}^{2n} \times \mathbb{R}^2$. For this purpose we consider a special symplectic immersion $\theta : C \to \mathbb{R}^2$ (this is known as Gromov's figure eight trick, see [G1], [AL]).

Exercise. (See Figure 3) Show that there exists a symplectic immersion $\theta : C \to \mathbb{R}^2(p,q)$ with the following properties:

- θ takes the zero-section $\{r = 0\}$ to the figure eight curve with ears of equal area. Thus the closed form $\theta^* pdq - rdt$ is exact (it is closed because θ is a symplectomorphism).
- θ is an embedding outside the thin necks and glues the thin necks together.
- the area of the interior ears is arbitrarily small, say ε each.

Note now that

$$\text{area}\,(C) = \int_0^2 (a_+(t) - a_-(t))dt = 2\int_0^1 (\max_x H_t - \min_x H_t)dt + 4\varepsilon = 2l + 4\varepsilon.$$

Thus the image $\theta(C)$ can be enclosed by a disc B of area $2l + 10\varepsilon$ (we take 10ε to make up for the extra bits).

Step 4. Consider now the symplectic immersion

$$\theta' = \text{id} \times \theta : \mathbb{R}^{2n} \times C \to \mathbb{R}^{2n} \times \mathbb{R}^2.$$

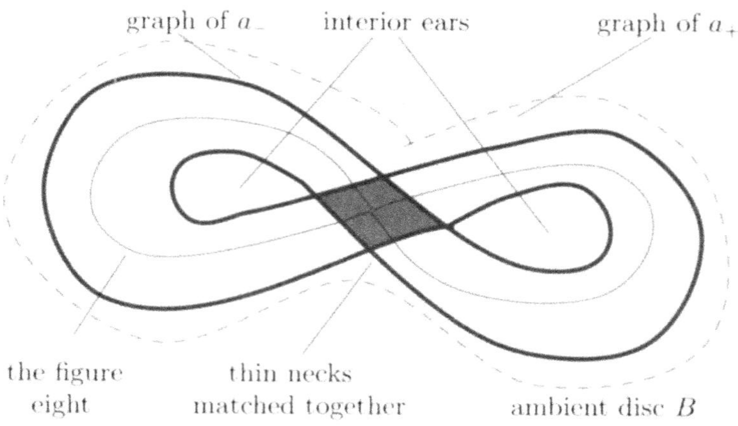

Figure 3

Obviously $\theta'(L') \subset \mathbb{R}^{2n+2}$ is an immersed Lagrangian submanifold and $\theta'(L') \subset \mathbb{R}^{2n} \times B$. We will show that $L'' = \theta'(L')$ is embedded.

Observe that the only place where double points could occur is in the thin necks. But for $t \in [-\varepsilon,\varepsilon]$, $g_t(L) = L$ and for $t \in [1-\varepsilon,1+\varepsilon]$, $g_t(L) = h_1(L)$ and by assumption $h_1(L) \cap L = \emptyset$ so there are no double points and θ' is an embedding.

Step 5. The only thing we are left to show is that L'' is rational so that we can apply 3.2.B. We will prove that $\gamma(L) = \gamma(L'')$. Let ϕ be the composition of the Lagrangian suspension and θ', that is the map

$$\phi : L \times S^1 \to \mathbb{R}^{2n} \times T^* S^1 \to \mathbb{R}^{2n}(p_1,\ldots,p_n,q_1,\ldots,q_n) \times \mathbb{R}^2(p,q)$$

sending (x,t) to $(g_t(x), \theta(-G(g_t(x),t),t))$. Then L'' is the image of $L \times S^1$ under ϕ. The group $H_1(L'')$ is generated by the cycles of the form $\phi(b)$ where either $b \subset L \times \{0\}$, or $b = \{x_0\} \times S^1$ for $x_0 \in L$. In the first case $\phi(b) = b \times \{\theta(0,0)\}$, so the symplectic areas of b and $\phi(b)$ coincide. In the second case write α for the orbit $\{g_t x_0\}$, $t \in [0;2]$. Note that

$$\int_b \phi^*(p_1 dq_1 + \cdots + p_n dq_n + p\,dq) = \int_\alpha p_1 dq_1 + \cdots + p_n dq_n + \int \theta^* p\,dq$$

$$= 0 + \int r\,dt = -\int_0^2 G(g_t(x_0),t)\,dt = 0.$$

This proves that L'' is a rational Lagrangian submanifold with $\gamma(L) = \gamma(L'')$. Since L'' is contained in $\mathbb{R}^{2n} \times B$ we get in view of 3.2.B that

$$\gamma(L'') \leq \text{area}\,(B) = 2l + 10\varepsilon,$$

for all $\varepsilon > 0$. Therefore $e(L) \geq \frac{1}{2}\gamma(L)$. $\qquad\square$

Chapter 4

The $\bar{\partial}$-Equation with Lagrangian Boundary Conditions

In this chapter we prove Theorem 3.2.B which states that $\gamma(L) \leq \pi r^2$ for any closed rational Lagrangian submanifold $L \subset B^2(r) \times \mathbb{R}^{2n-2}$. The proof is based on Gromov's techniques of pseudo-holomorphic discs.

4.1 Introducing the $\bar{\partial}$-operator

Identify $\mathbb{R}^{2n}(p_1, q_1, \ldots, p_n, q_n)$ with

$$\mathbb{C}^n(p_1 + iq_1, \ldots, p_n + iq_n) = \mathbb{C}^n(w_1, \ldots, w_n).$$

Denote by $\langle \, , \, \rangle$ the Euclidean scalar product. The three geometric structures we get in this way are the Euclidean, the symplectic and the complex structure. They are related by the following formula

$$\langle \xi, \eta \rangle = \omega(\xi, i\eta).$$

We will check this formula in the case $n = 1$. Let $\xi = (p', q')$ and $\eta = (p'', q'')$. Then

$$dp \wedge dq(\xi, i\eta) = dp \wedge dq\left(\begin{pmatrix} p' \\ q' \end{pmatrix}, \begin{pmatrix} -q'' \\ p'' \end{pmatrix} \right) = p'p'' + q'q'' = \langle \xi, \eta \rangle.$$

In what follows we will measure areas and lengths using the Euclidean metric. Consider the unit disc $D^2 \subset \mathbb{C}$ with coordinate $z = x + iy$. For a smooth map $f : D^2 \to \mathbb{C}^n$ we define the $\bar{\partial}$-operator, $\bar{\partial} : C^\infty(D^2, \mathbb{C}^n) \to C^\infty(D^2, \mathbb{C}^n)$ by

$$\bar{\partial} f = \frac{1}{2}\left(\frac{\partial f}{\partial x} + i\frac{\partial f}{\partial y} \right).$$

Example. Let $f : \mathbb{C} \to \mathbb{C}, z \mapsto \bar{z}$. So $f(x,y) = x - iy$ and $\bar{\partial} f = \frac{1}{2}(1+1) = 1$. We observe that $\bar{\partial} f = \frac{\partial f}{\partial \bar{z}}$.

Let us introduce two useful geometric quantities associated with a map $f : D^2 \to \mathbb{C}^n$. The *symplectic area* of f is given by

$$\omega(f) = \int_{D^2} f^* \omega$$

and the *Euclidean area* of f is given by

$$\text{Area } (f) = \int_{D^2} \sqrt{\left\langle \frac{\partial f}{\partial x}, \frac{\partial f}{\partial x} \right\rangle \left\langle \frac{\partial f}{\partial y}, \frac{\partial f}{\partial y} \right\rangle - \left\langle \frac{\partial f}{\partial x}, \frac{\partial f}{\partial y} \right\rangle^2} \, dxdy.$$

Proposition 4.1.A.

   i) $\text{Area } (f) \leq 2 \int_{D^2} |\bar{\partial} f|^2 dxdy + \omega(f),$

   ii) $\text{Area } (f) \geq |\omega(f)|.$

Proof. Given $\xi, \eta \in \mathbb{C}^n$ we have the following inequality:

$$\sqrt{|\xi|^2|\eta|^2 - \langle \xi, \eta \rangle^2} \leq |\xi||\eta| \leq \frac{1}{2}(|\xi|^2 + |\eta|^2).$$

But

$$\frac{1}{2}|\xi + i\eta|^2 + \omega(\xi, \eta) = \frac{1}{2}(|\xi|^2 + |\eta|^2) + \langle \xi, i\eta \rangle + \langle \xi, -i\eta \rangle = \frac{1}{2}(|\xi|^2 + |\eta|^2),$$

so we get

$$\sqrt{|\xi|^2|\eta|^2 - \langle \xi, \eta \rangle^2} \leq \frac{1}{2}|\xi + i\eta|^2 + \omega(\xi, \eta).$$

Integrating this pointwise inequality we get 4.1.A.i.

To prove 4.1.A.ii it suffices again to verify the pointwise inequality. Assume that $\eta \neq 0$ and note that $\langle \eta, i\eta \rangle = 0$. Projecting ξ on η and $i\eta$ we get

$$\left\langle \xi, \frac{\eta}{|\eta|} \right\rangle^2 + \left\langle \xi, \frac{i\eta}{|i\eta|} \right\rangle^2 \leq |\xi|^2.$$

Since $|\eta| = |i\eta|$ the above reads

$$\langle \xi, \eta \rangle^2 + \omega(\xi, \eta)^2 \leq |\xi|^2|\eta|^2,$$

hence

$$|\omega(\xi, \eta)| \leq \sqrt{|\xi|^2|\eta|^2 - \langle \xi, \eta \rangle^2}. \qquad \square$$

4.2 The boundary value problem

Let $L \subset \mathbb{C}^n$ be a closed Lagrangian submanifold and let $g : D^2 \times \mathbb{C}^n \to \mathbb{C}^n$ be a smooth map which is bounded together with all its derivatives. Fix a class $\alpha \in H_2(\mathbb{C}^n, L)$. Given this data consider the following problem.

Find a smooth map $f : (D^2, \partial D^2) \to (\mathbb{C}^n, L)$ *such that*

$$\left\{ \begin{array}{l} \bar{\partial} f(z) = g(z, f(z)) \\ [f] = \alpha. \end{array} \right. \qquad (P(\alpha, g))$$

Example. If $g = 0, \alpha = 0$ then the space of solutions of $P(0,0)$ consists of the constant mappings $f(z) \equiv w$ for $w \in L$. To see this first observe that $\omega(f) = 0$. Indeed, since $\alpha = 0$ and L is Lagrangian the curve $f(\partial D^2)$ bounds a 2-chain in L with zero symplectic area. This chain together with $f(D^2)$ forms a closed surface in $\mathbb{C}^n$. Since ω is exact the symplectic area of this surface vanishes. Therefore $\omega(f) = 0$. Further, since $g = 0$ we get that $\bar{\partial} f = 0$. So the first part of 4.1.A yields Area $(f) = 0$ and hence $\frac{\partial f}{\partial x}$ is parallel to $\frac{\partial f}{\partial y}$. On the other hand $\frac{\partial f}{\partial x} = -i \frac{\partial f}{\partial y}$ and therefore $\frac{\partial f}{\partial x} \perp \frac{\partial f}{\partial y}$. Consequently $\frac{\partial f}{\partial x} = \frac{\partial f}{\partial y} = 0$. So f is a constant map and because of the boundary condition its image lies in L.

Assume now that we have a sequence of functions $\{g_n\}$ which C^∞-converges to some function g. Let f_n be the solutions of the corresponding problems $P(\alpha, g_n)$. Gromov's famous compactness theorem (see [G1], [AL]) states that either $\{f_n\}$ contains a subsequence that converges to a solution of $P(\alpha, g)$ or bubbling off takes place. In order to explain what bubbling off is we introduce the concept of a cusp solution.

Definition. Consider the following data:

- A decomposition $\alpha = \alpha' + \beta_1 + \cdots + \beta_k$, where $\beta_j \neq 0$, $j = 1, \ldots k$.
- A solution f of $P(\alpha', g)$.
- Solutions h_j of $P(\beta_j, 0)$, these are the so called pseudo-holomorphic discs.

This object is called a *cusp solution* of $P(\alpha, g)$ and $f(D^2) \cup h_1(D^2) \cup \cdots \cup h_k(D^2)$ is called its image.

The bubbling off phenomenon means that there exists a subsequence of $\{f_n\}$ (which we denote again by $\{f_n\}$) which converges to a cusp solution of $P(\alpha, g)$. The only feature of this convergence which is important for our purposes is the continuity of the Euclidean area:

$$\text{Area } (f_n) \to \text{Area } (f) + \sum_{j=1}^{k} \text{Area } (h_j).$$

The complete definition of the convergence is quite sophisticated (see [G1], [AL]), and we omit it. An illustrating example will be given in 4.4 below.

Using the compactness theorem, Gromov established the following important result [G1].

Persistence Principle. *Consider a "generic" family $g_s(z, w)$, $s \in [0, 1]$ with $g_0 = 0$. Then either $P(0, g_s)$ has a solution for all s or bubbling off occurs at some $s_\infty \leq 1$, that is there exists a subsequence $s_j \to s_\infty$ such that the sequence of solutions of $P(0, g_{s_j})$ converges to a cusp solution of $P(0, g_{s_\infty})$.*

The word "generic" should be interpreted as follows. One can endow the space of all families g_s with an appropriate Banach manifold structure. Generic families form a residual subset (that is a countable intersection of open and dense subsets) in this space. In particular, every family g_s becomes generic after an arbitrarily small perturbation. We refer to [G1], [AL] for further details.

4.3   An application to the Liouville class

We present here Sikorav's proof of Theorem 3.2.B [S1]. Suppose that $L \subset B^2(r) \times \mathbb{C}^{n-1}$ is a closed Lagrangian submanifold. Take $g(z, w) = (\sigma, 0, \ldots, 0) \in \mathbb{C}^n$ for some $\sigma \in \mathbb{C}$.

Lemma 4.3.A. *If $|\sigma| > r$ then $P(0, g)$ has no solutions.*

Proof. Suppose that f is a solution. Denote by ϕ its first (complex) coordinate. Thus

$$\frac{\partial \phi}{\partial x} + i\frac{\partial \phi}{\partial y} = 2\sigma.$$

Since $L \subset B^2(r) \times \mathbb{C}^{n-1}$, we have that $\left|\phi|_{\partial D^2}\right| \leq r$. Note that

$$2\pi\sigma = \int_{D^2} \frac{\partial \phi}{\partial x} + i\frac{\partial \phi}{\partial y} \, dxdy$$

$$= \int_{D^2} d(\phi dy - i\phi dx)$$

$$= \int_{S^1} \phi dy - i\phi dx.$$

Now we can write $x + iy = e^{2\pi it}$ and $dx + idy = 2\pi ie^{2\pi it}dt$, so $dy - idx = 2\pi e^{2\pi it}dt$. Therefore

$$2\pi|\sigma| = 2\pi\left|\int_0^1 e^{2\pi it}\phi(e^{2\pi it})dt\right| \leq 2\pi r$$

and hence $|\sigma| \leq r$. $\square$

Take now any σ with $|\sigma| > r$ and apply the persistence principle to the family $g_s = (s\sigma, 0, \ldots, 0)$, $s \in [0, 1]$. The previous lemma tells us that there is no solution for $s = 1$, so we have that for a small perturbation of g_s bubbling off takes place. For the sake of simplicity we assume that it happens in g_s itself. The general argument goes through without changes (make estimates up to ε) and is left to the reader.

So we have a sequence $s_n \to s_\infty \leq 1$ and a decomposition $0 = \alpha + \beta_1 + \cdots + \beta_k$, $\beta_j \neq 0$. Let f_n be the solutions of $P(0, g_{s_n})$, f_∞ a solution of $P(\alpha, g_{s_\infty})$ and $h_1, \ldots h_k$ holomorphic discs with $[h_j] = \beta_j$ satisfying

$$\text{Area } (f_n) \to \text{Area } (f_\infty) + \sum_{j=1}^{k} \text{Area } (h_j).$$

Applying both parts of 4.1.A and using the fact that the discs h_j are holomorphic we get that Area $(h_j) = \omega(h_j) \geq \gamma(L)$. This inequality follows from the fact that $[h_j] = \beta_j \neq 0$. From 4.1.A.ii we deduce that

$$\text{Area } (f_\infty) \geq |\omega(f_\infty)| = \left| \sum \omega(h_j) \right| \geq \gamma(L).$$

Thus Area $(f_\infty) + \sum \text{Area } (h_j) \geq 2\gamma(L)$. On the other hand 4.1.A.i implies that

$$\text{Area } (f_n) \leq 2\pi s_n^2 |\sigma|^2 \leq 2\pi |\sigma|^2.$$

We use here that $\omega(f_n) = 0$ (since $[f_n] = 0$) and $\bar{\partial} f_n = g_{s_n}$. Putting these two inequalities together we get $2\pi |\sigma|^2 \geq 2\gamma(L)$. This is true for all σ with $|\sigma| > r$ so we have $\pi r^2 \geq \gamma(L)$, which proves the theorem. $\qquad\square$

Proof of 3.2.A. Consider a closed Lagrangian submanifold $L \subset B^2(r) \times \mathbb{C}^{n-1}$. Then 4.3.A above implies that the problem

$$\begin{cases} \bar{\partial} f(z) = (s\sigma, 0, \ldots, 0), & |\sigma| > r \\ [f] = 0. \end{cases}$$

has no solution for $s = 1$. The persistence principle implies that bubbling off must take place. This means that there exists a non-zero class β_1 which is represented by a holomorphic disc h_1. Since $h_1 \neq$ constant, we get $\omega(h_1) > 0$. Since we found a disc in $\mathbb{C}^n$ spanned by $h_1(\partial D^2)$ which has non-zero symplectic area, we conclude that $\lambda_L \neq 0$. $\qquad\square$

4.4   An example

In this section we sketch an elementary example where all the phenomena appearing in the proof can be seen explicitly. Let $L = \partial D^2 \subset \mathbb{C}$ and let $\sigma = 1$.

We wish to find all maps $f : D^2 \to \mathbb{C}$ such that $f(\partial D^2) \subset \partial D^2$ and

$$\left\{ \begin{array}{l} \bar\partial f(z, \bar z) = s \\ [f|_{\partial D^2}] = 0. \end{array} \right. \tag{4.4.A}$$

Since $\frac{\partial f}{\partial \bar z} = s$ then $f(z, \bar z) = s\bar z + u(z)$ where u is a holomorphic function on D^2. We claim that the function $s + zu(z)$ is holomorphic, takes ∂D^2 to ∂D^2 and that $(s + zu(z))|_{\partial D^2}$ has degree 1. Indeed,

$$zf(z, \bar z) = s|z|^2 + zu(z)$$

so

$$|z|\,|f(z, \bar z)| = \left| s|z|^2 + zu(z) \right|.$$

If $|z| = 1$ this reduces to $|f(z, \bar z)| = |s + zu(z)|$ but $|f(z, \bar z)| = 1$. So $s + zu(z)$ is a holomorphic function taking ∂D^2 to ∂D^2. Observe that $\deg f = 0$ and $\deg z = 1$ so the total degree, $\deg zf = 1$ and hence $\deg (s + zu(z)) = 1$. All such holomorphic functions are known as isometries of the hyperbolic metric in the unit disc. They have the form

$$e^{i\theta} \frac{1 - \bar\alpha z}{z - \alpha}$$

for $\theta \in \mathbb{R}$ and $|\alpha| > 1$. Thus $s + zu(z) = e^{i\theta} \frac{1 - \bar\alpha z}{z - \alpha}$ so

$$zu(z) = \frac{e^{i\theta} + \alpha s - z(s + e^{i\theta}\bar\alpha)}{z - \alpha}.$$

Since u is holomorphic it cannot have any poles, so $e^{i\theta} + \alpha s = 0$ and hence $\alpha = -\frac{e^{i\theta}}{s}$. Now $1 < |\alpha| = |\frac{1}{s}|$ implies that $s < 1$ and that there are no solutions of 4.4.A for $s \geq 1$. So bubbling off must occur at $s = 1$. For the sake of simplicity we put $\theta = 0$. Then

$$u(z) = -\frac{s - \frac{1}{s}}{z + \frac{1}{s}} = \frac{1 - s^2}{sz + 1}$$

so $f_s(z, \bar z) = s\bar z + \frac{1 - s^2}{sz + 1}$. When s goes to 1, $f_s(z, \bar z) \to \bar z$ for all $z \neq -1$ and this convergence is uniform outside every neighbourhood of -1.

Consider the graphs of f_s in $D^2 \times D^2 \subset \mathbb{C} \times \mathbb{C}$. Set $w = f_s(z, \bar z)$ so

$$(w - s\bar z)(sz + 1) = 1 - s^2.$$

When $s \to 1$ this equation goes to $(w - \bar z)(z + 1) = 0$ and the graph becomes the union of two curves

$$w = \bar z \quad \text{and}$$
$$z = -1.$$

Here $w = \bar{z}$ is the graph of f_∞ and $\{z = -1\}$ corresponds to the holomorphic disc with boundary on $\{-1\} \times L$. Projecting onto the w-coordinate we get bubbling off. Indeed, $f_\infty(z) = \bar{z}$ is a solution of $P(-a, 1)$ where $a = [S^1]$ and the holomorphic disc $h(z) = z$ is a solution of $P(a, 0)$.

In order to visualize the bubbling off phenomenon we restrict to the real axes. Consider the graphs of the corresponding functions $f_s(x) = sx + \frac{1-s^2}{sx+1}$ for $x \in [-1, 1]$. We get the following picture, see Figure 4. The graphs of f_s converge to the union of two curves, the graph of the real part of f_∞ and the segment $I = [-1, 1]$ which is the real part of the holomorphic disc $\{-1\} \times D^2$.

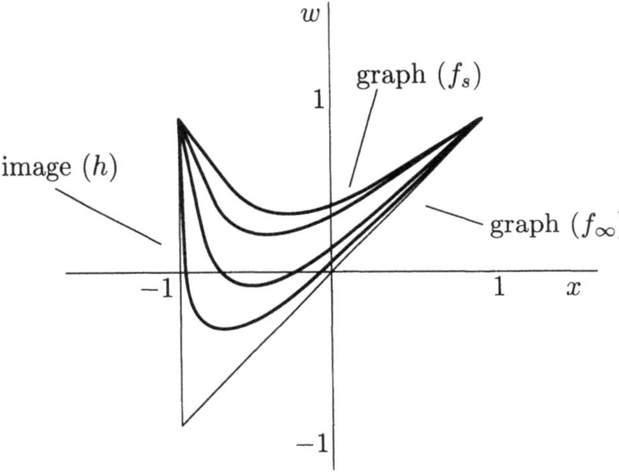

Figure 4

Chapter 5

Linearization of Hofer's Geometry

In this chapter we interpret the function $\rho(\mathbb{1}, \phi)$ as the distance between a point and a subset in a linear normed space. Later on this will enable us in some interesting cases to find lower bounds for $\rho(\mathbb{1}, \phi)$.

5.1 The space of periodic Hamiltonians

Denote by $\mathcal{F}$ the space of all smooth normalized Hamiltonian functions $F : M \times \mathbb{R} \to \mathbb{R}$ which are 1-periodic in time: $F(x, t+1) = F(x, t)$ for all $x \in M$. We will often consider such F as functions on $M \times S^1$, where $S^1 = \mathbb{R}/\mathbb{Z}$. For a function $F \in \mathcal{F}$ denote by ϕ_F the time-one-map f_1 of the corresponding Hamiltonian flow $\{f_t\}$.

Note that every Hamiltonian diffeomorphism ϕ can be represented in this way. Indeed, let $\{g_t\}$ be any flow with $g_1 = \phi$. Take a function $a : [0; 1] \to [0; 1]$ such that $a \equiv 0$ near 0 and $a \equiv 1$ near 1. Consider the new flow $f_t = g_{a(t)}$ and extend it on the whole $\mathbb{R}$ by the formula $f_{t+1} = f_t f_1$. Clearly we get a smooth flow. The claim follows from the next exercise.

Exercise 5.1.A. Prove that a flow $\{f_t\}$, $t \in \mathbb{R}$ is generated by a function from $\mathcal{F}$ if and only if $f_{t+1} = f_t f_1$ for all t.

Consider a subset $\mathcal{H} \subset \mathcal{F}$ defined by

$$\mathcal{H} = \{H \in \mathcal{F} \mid \phi_H = \mathbb{1}\}.$$

In other words Hamiltonians from $\mathcal{H}$ generate *loops* of Hamiltonian diffeomorphisms (or *Hamiltonian loops*). Introduce a norm on $\mathcal{F}$ as follows:

$$|||F||| = \max_t ||F_t|| = \max_t (\max_x F(x, t) - \min_x F(x, t)).$$

Now we are ready to present the main theorem of this chapter.

Theorem 5.1.B. *For every $F \in \mathcal{F}$*

$$\rho(\mathbb{1}, \phi_F) = \inf_{H \in \mathcal{H}} |||F - H|||.$$

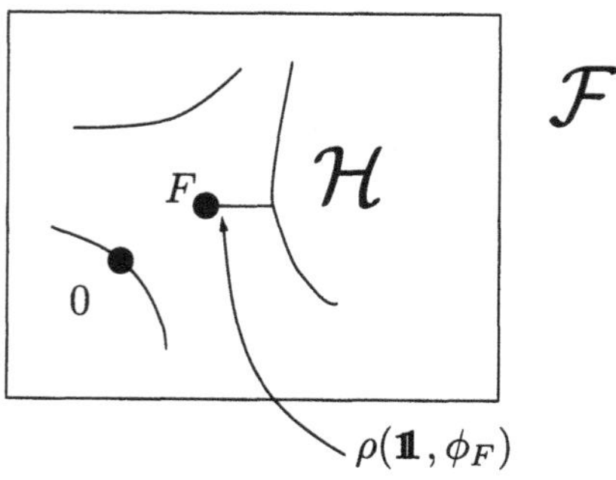

Figure 5

Note that the right-hand side is simply the distance from F to $\mathcal{H}$ in the sense of our norm (see Figure 5). Thus the set $\mathcal{H}$ carries a lot of information about Hofer's geometry. In the next chapters we will establish some interesting properties of $\mathcal{H}$ and will look closely at Hamiltonian loops.

Theorem 5.1.B is a simple consequence of the following fact.

Lemma 5.1.C. *For every $\phi \in \mathrm{Ham}(M)$*

$$\rho(\mathbb{1}, \phi) = \inf |||F|||,$$

where the infimum is taken over all Hamiltonians from $\mathcal{F}$ which generate ϕ.

In the terminology of some of my papers this means that the "coarse" Hofer's norm coincides with the usual one.

Proof of 5.1.C. For $\phi \in \mathrm{Ham}(M, \Omega)$ set $r(\mathbb{1}, \phi) = \inf |||F|||$ where F runs over all Hamiltonians $F \in \mathcal{F}$ which generate ϕ. Clearly $r(\mathbb{1}, \phi) \geq \rho(\mathbb{1}, \phi)$. Our task is to prove the converse inequality. Fix a positive number ϵ. Choose a path $\{f_t\}, t \in [0; 1]$ of Hamiltonian diffeomorphisms such that $f_0 = \mathbb{1}, f_1 = \phi$ and such that $\int_0^1 m(t)dt \leq \rho(\mathbb{1}, \phi) + \epsilon$ where $m(t) = ||F_t||$. We can assume without loss of generality that $F \in \mathcal{F}$ and that $m(t) > 0$ for all t. Indeed, to guarantee the periodicity one makes a time reparametrization as it is explained in the beginning of this section. The justification of the second assumption is given in the next section. Denote by $\mathcal{C}$ the space of all C^1-smooth orientation preserving diffeomorphisms of S^1 which fix 0. Note that for $a \in \mathcal{C}$ the path $f_a = \{f_{a(t)}\}$ is generated by the normalized Hamiltonian function $F^a(x, t) = a'(t)F(x, a(t))$, where a' denotes the derivative with respect to t (see 1.4.A above). Take now $a(t)$ as the inverse of

$$b(t) = \frac{\int_0^t m(s)ds}{\int_0^1 m(s)ds}.$$

Note that

$$|||F^a||| = \max_t a'(t)m(a(t)) = \max_t (m(t)/b'(t)) = \int_0^1 m(t)dt.$$

We conclude that $|||F^a||| \leq \rho(\mathbb{1},\phi)+\epsilon$. Approximating a in the C^1-topology by a smooth diffeomorphism from $\mathcal{C}$ we see that one can find a smooth normalized Hamiltonian, say $\tilde{F}$, which generates ϕ and such that $|||\tilde{F}||| \leq \rho(\mathbb{1},\phi) + 2\epsilon$. Since this can be done for an arbitrary ϵ we conclude that $r(\mathbb{1},\phi) \leq \rho(\mathbb{1},\phi)$. This completes the proof. $\qquad\square$

Proof of 5.1.B. Write $\{f_t\}$ for the Hamiltonian flow generated by F. Let $\{g_t\}$ be any other Hamiltonian flow generated by $G \in \mathcal{F}$ with $g_1 = \phi_F$. Decompose g_t as $h_t \circ f_t$. It follows from 5.1.A that $\{h_t\}$ is a loop of Hamiltonian diffeomorphisms, that is its normalized Hamiltonian H belongs to $\mathcal{F}$ and $h_0 = h_1 = \mathbb{1}$. Vice versa, for every loop $\{h_t\}$ the flow $\{h_t \circ f_t\}$ is generated by a Hamiltonian from $\mathcal{F}$,[1] and its time one map equals ϕ_F. Further,

$$G(x,t) = H(x,t) + F(h_t^{-1}x,t).$$

Set $H'(x,t) = -H(h_t x,t)$. Note that H' generates a loop $\{h_t^{-1}\}$, and thus belongs to $\mathcal{H}$. On the other hand the expression for G above implies that $|||G||| = |||F - H'|||$. In view of the discussion above to each G corresponds unique H' and vice versa. Thus the required statement follows immediately from Lemma 5.1.C. $\qquad\square$

5.2 Regularization

Here we justify the assumption $m(t) > 0$ made in the proof of Lemma 5.1.C. A flow $\{f_t\}$ is called *regular* if for every t the normalized Hamiltonian function F_t does not vanish identically. In other words, for every t the tangent vector to the path $\{f_t\}$ does not vanish.

---

[1]Note that in general the flow $\{f_t \circ h_t\}$ (mind the order) is not generated by a periodic Hamiltonian!

Proposition 5.2.A. *Let $\{f_t\}$ be a flow generated by a Hamiltonian from F. Then there exists an arbitrary small (in the C^∞-sense) loop $\{h_t\}$ such that the flow $\{h_t^{-1}f_t\}$ is regular.*

The proof is divided in several steps.

1) Let us analyze the problem. Assume that $\{h_t\}$ is a loop generated by a Hamiltonian $H \in \mathcal{H}$. Then the Hamiltonian of the flow $\{h_t^{-1}f_t\}$ is given by $-H(h_t x, t) + F(h_t x, t)$. We have to prove that for every t this expression does not vanish identically. But this is equivalent to the assertion that

$$F(x,t) - H(x,t) \not\equiv 0 \qquad (5.2.\text{B})$$

for all t. So our task is to produce an arbitrary small $H \in \mathcal{H}$ which satisfies 5.2.B.

2) Let us introduce another useful notion. *A k-parameter variation of the constant loop is a smooth family of loops $\{h_t(\epsilon)\}$*, where ϵ belongs to a neighbourhood of 0 in $\mathbb{R}^k$ and $h_t(0) = \mathbb{1}$ for all t. When M is open we assume in addition that the supports of all $h_t(\epsilon)$ are contained in some compact subset of M.

Here is a convenient way to produce variations. Let us start with the 1-parameter case. Take a function $G \in \mathcal{F}$ such that

$$\int_0^1 G(x,t)dt = 0 \qquad (5.2.\text{C})$$

for every $x \in M$. Then define $h_t(\epsilon) \in \mathrm{Ham}(M,\Omega)$ as the time-ϵ-map of the Hamiltonian flow generated by the time-independent Hamiltonian function $\int_0^t G(x,s)ds$.

Exercise 5.2.D. Let $H(x,t,\epsilon)$ be the normalized Hamiltonian function of the loop $\{h_t(\epsilon)\}$. Show that

$$\frac{\partial}{\partial \epsilon}\Big|_{\epsilon=0} H(x,t,\epsilon) = G(x,t).$$

It is natural to construct k-parameter variations as compositions of 1-parameter variations:

$$h_t(\epsilon_1,\ldots,\epsilon_k) = h_t^{(1)}(\epsilon_1) \circ \cdots \circ h_t^{(k)}(\epsilon_k).$$

Here every $h^{(j)}$ is constructed with the help of a function $G^{(j)}$ as above. Exercise 5.2.D yields that the partial derivative of the Hamiltonian $H(x,t,\epsilon)$ with respect to ϵ_j at $\epsilon = 0$ equals $G^{(j)}$.

3) Fix a point $y \in M$, and consider the $2n$-dimensional linear space $E = T_y^* M$. Choose $2n$ smooth closed curves $\alpha_1(t),\ldots,\alpha_{2n}(t)$ (where $t \in S^1$) which satisfy the following conditions:

- $\int_0^1 \alpha_j(t)\,dt = 0$ for all $j = 1, \ldots, 2n$;
- the vectors $\alpha_1(t), \ldots, \alpha_{2n}(t)$ are linearly independent for every t.

Here is a construction of such a system of curves. Choose a basis u_1, v_1, $\ldots$, u_n, v_n in E and take curves of the form $u_j \cos 2\pi t + v_j \sin 2\pi t$ and $-u_j \sin 2\pi t + v_j \cos 2\pi t$.

4) Choose now functions $G_1(x, t), \ldots, G_{2n}(x, t)$ from $\mathcal{F}$ which satisfy condition 5.2.C above and such that $d_y G_t^{(j)} = \alpha_j(t)$. Take the corresponding $2n$-parameter variation $\{h_t(\epsilon)\}$ of the constant loop as in step 2. Consider the mapping $\Phi : S^1 \times \mathbb{R}^{2n}(\epsilon_1, \ldots, \epsilon_{2n}) \to E$ defined by

$$(t, \epsilon) \to d_y(F_t - H_t(\epsilon)).$$

It follows that Φ is a submersion in some neighbourhood U of the circle $\{\epsilon = 0\}$. Indeed, our construction together with the discussion in step 2 imply that

$$\frac{\partial}{\partial \epsilon_j}\Big|_{\epsilon=0} \Phi(t, \epsilon) = \alpha_j(t).$$

But these vectors generate the whole E. Denote by Ψ the restriction of Φ to $S^1 \times U$. Since Ψ is a submersion, the set $\Psi^{-1}(0)$ is a one-dimensional submanifold of $S^1 \times U$, so its projection to U is nowhere dense. Thus there exist arbitrary small values of the parameter ϵ such that $d_y(F_t - H_t(\epsilon)) \neq 0$ for all t. Therefore for every t the condition 5.2.B above is satisfied. This completes the argument. $\qquad\square$

5.3 Paths in a given homotopy class

A *homotopy* is simply a smooth 1-parameter family of paths. We will consider homotopies of non-closed paths with fixed end points, as well as homotopies of loops based at $\mathbb{1}$, unless otherwise stated. When the manifold M is open we as usually assume that supports of all diffeomorphisms appearing in 2-parameter families are contained in an ambient compact subset of M.

Take a function $F \in \mathcal{F}$, and denote by $\{f_t\}$ the corresponding Hamiltonian flow. Consider the quantity

$$l(F) = \inf \operatorname{length}\{g_t\},$$

where the infimum is taken over all Hamiltonian paths $\{g_t\}$, $t \in [0; 1]$ with $g_0 = \mathbb{1}$, $g_1 = \phi_F$ which are homotopic to $\{f_t\}$ with fixed end points. Let us sketch a useful interpretation of this quantity. Consider the universal cover Z of $(\operatorname{Ham}(M, \Omega), \mathbb{1})$. It is defined in the standard manner with the minor exception that we consider smooth paths and smooth homotopies only. The Finsler structure on $\mathrm{Ham}(M, \Omega)$ lifts canonically to the universal cover. This gives rise to the notion of the length of a smooth curve on Z, and therefore to the distance function $\tilde{\rho}$ on Z. Denote by $\tilde{1\!\!1}$ the canonical lift of $1\!\!1$ to Z, and by $\tilde{\phi}_F$ the lift of ϕ_F associated to the path $\{f_t\}$, $t \in [0; 1]$. With this language $l(F) = \tilde{\rho}(\tilde{1\!\!1}, \tilde{\phi}_F)$, so this quantity is responsible for the geometry of the universal cover.

Denote by $\mathcal{H}_c$ the set of all Hamiltonians from $\mathcal{H}$ which generate contractible loops. In other words $\mathcal{H}_c$ is the path connected component of 0 in $\mathcal{H}$.

Theorem 5.3.A. *For every $F \in \mathcal{F}$*

$$l(F) = \inf_{H \in \mathcal{H}_c} |||F - H|||.$$

This can be proved exactly in the same way as Theorem 5.1.B above. In the course of the proof one should take into account the following simple additional observations:

- Time reparametrizations as well as the regularization procedure 5.2 do not change the homotopy class of a path with fixed end points. Thus one reduces the problem to the minimization of $|||G|||$ over all $G \in \mathcal{F}$ with $\phi_G = \phi_F$ whose Hamiltonian flow is homotopic to $\{f_t\}$ (cf. 5.1.C above).
- If $g_t = h_t \circ f_t$, where $\{f_t\}$ and $\{g_t\}$ are homotopic with fixed end points, then the loop h_t is contractible (cf. the final argument in the proof of 5.1.B).

The details of the proof are left to the reader.

Chapter 6

Lagrangian Intersections

The theory of Lagrangian intersections studies one of the most surprising phenomena in symplectic topology. In this chapter we review some results from this theory which being combined with the linearization idea above give a rather powerful tool for investigation of the geometry of the group of Hamiltonian diffeomorphisms.

6.1 Exact Lagrangian isotopies

Let (V^{2n}, ω) be a symplectic manifold and let N^n be a closed manifold. Let

$$\Phi : N \times [0, 1] \to V$$

be a smooth family of Lagrangian embeddings i.e. Φ is a Lagrangian isotopy. Note that $\Phi^*\omega$ must be of the form $\alpha_s \wedge ds$ where $\{\alpha_s\}$ is a family of 1-forms on N (since $\Phi^*\omega$ vanishes on the fibers $N \times \{\text{point}\}$). Furthermore, observe that $d\Phi^*\omega = d\alpha_s \wedge ds = 0$ which implies that α_s is closed for all s.

Definition. A Lagrangian isotopy Φ is *exact* if α_s is exact for all s.

Exercise 6.1.A. Show that a Lagrangian isotopy is exact if and only if it can be extended to an ambient Hamiltonian isotopy of V. *Hint:* Write $\alpha_s = dH_s$ on N and extend $H_s \circ \Phi_s^{-1}$ to a time dependent normalized Hamiltonian function on V.

Example. Let V be a surface and $N = S^1$. Then Φ is exact if and only if the oriented area between $\Phi(N \times \{0\})$ and $\Phi(N \times \{s\})$ vanishes for all s. See Figure 6 for the case $V = S^2$, and Figure 7 for the case $V = T^*S^1 = \mathbb{R} \times S^1$. Note that in the case of the cylinder it is possible to find a symplectic isotopy describing the right-hand side picture.

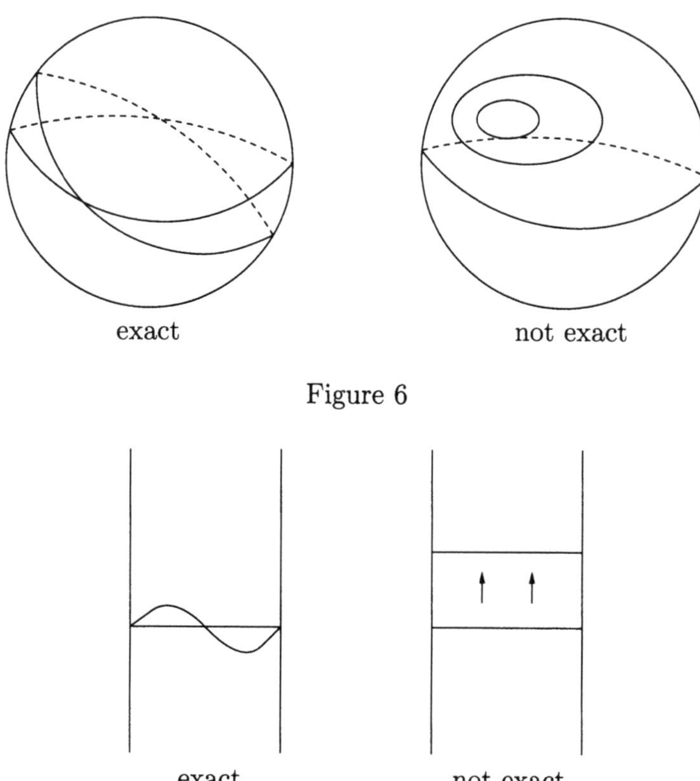

Figure 6

Figure 7

The next result plays an important role in our study of Hofer's geometry. Assume that $\{h_t\}$ is a loop of Hamiltonian diffeomorphisms of (M, Ω) generated by $H \in \mathcal{H}$. Let $L \subset M$ be a closed Lagrangian submanifold. Consider the Lagrangian suspension (see 3.1.E above).

$$L \times S^1 \to (M \times T^*S^1, \Omega + dr \wedge dt),$$

$$(x, t) \mapsto (h_t x, -H(h_t x, t), t).$$

Our aim is to investigate the behaviour of this Lagrangian embedding under a 1-parameter deformation. Let $\{h_{t,s}\}$, $s \in [0; 1]$ be a smooth family of Hamiltonian loops. Denote by

$$\Phi : L \times S^1 \times [0; 1] \to M \times T^*S^1$$

the corresponding family of Lagrangian suspensions.

Theorem 6.1.B. *The Lagrangian isotopy Φ is exact.*

In other words, a homotopy of Hamiltonian loops gives rise to an exact Lagrangian isotopy of Lagrangian suspensions. The proof of the theorem is based on the following auxiliary result. Denote by $H(x, t, s)$ the normalized Hamiltonian of the loop $\{h_{t,s}\}$.

Proposition 6.1.C. *For every $x \in M$ and $s \in [0; 1]$*

$$\int_0^1 \frac{\partial H}{\partial s}(h_{t,s}x, t, s)dt = 0.$$

Proof of the Proposition. The proof is based on the following formula, which is valid for an arbitrary 2-parameter family of diffeomorphisms on a manifold. We leave its verification to the reader (see also [B1]). Consider vector fields $X_{t,s}$ and $Y_{t,s}$ on M such that

$$\frac{d}{dt}h_{t,s}x = X_{t,s}(h_{t,s}x), \quad \text{and} \quad \frac{d}{ds}h_{t,s}x = Y_{t,s}(h_{t,s}x).$$

Then

$$\frac{\partial}{\partial s}X_{t,s} = \frac{\partial}{\partial t}Y_{t,s} + [X_{t,s}, Y_{t,s}].$$

Note that for every t and s fields $X_{t,s}$ and $Y_{t,s}$ are Hamiltonian vector fields. Of course $X_{t,s} = \operatorname{sgrad} H_{t,s}$ where H is already introduced above. Write $Y_{t,s} = \operatorname{sgrad} F_{t,s}$. Recall that

$$[\operatorname{sgrad} H, \operatorname{sgrad} F] = -\operatorname{sgrad}\{H, F\}.$$

Thus we get that

$$\frac{\partial H_{t,s}}{\partial s} = \frac{\partial F_{t,s}}{\partial t} - \{H_{t,s}, F_{t,s}\} = \frac{\partial F_{t,s}}{\partial t} + dF_{t,s}(\operatorname{sgrad} H_{t,s}).$$

But the last expression evaluated at point $h_{t,s}x$ equals

$$\frac{d}{dt}F(h_{t,s}x, t).$$

We conclude that

$$\frac{\partial H_{t,s}}{\partial s}(h_{t,s}x)$$

is the full derivative of a periodic function, thus its integral over the period must vanish. This completes the proof. $\square$

Proof of 6.1.B. Write $\Phi^*(\Omega + dr \wedge dt)$ as $\alpha_s \wedge ds$. We have to check that α_s is an exact form. The form α_s can be calculated explicitly.

Exercise (cf. 3.1.E). Show that

$$\alpha_s(\xi) = \Omega(h_{t,s*}\xi, \frac{\partial h_{t,s}}{\partial s}x)$$

for all $x \in L$, $\xi \in T_xL$, and

$$\alpha_s(\frac{\partial}{\partial t}) = \frac{\partial H}{\partial s}(h_{t,s}x, t, s).$$

Note that the first homology group $H_1(L \times S^1, \mathbb{Z})$ is generated by split cycles of the form $C = \beta \times \{0\}$ and $D = \{y\} \times S^1$. Here β is a cycle on L, and y is a point of L. In order to prove that the form α_s is exact it suffices to verify that its integrals over all 1-cycles vanish. For the cycles of the form C this follows from the fact that $h_{0,s} \equiv \mathbb{1}$ for all s. Thus the exercise above implies that α_s vanishes on all vectors tangent to $L \times \{0\}$. Further,

$$\int_D \alpha_s = \int_0^1 \frac{\partial H}{\partial s}(h_{t,s}y, t, s)dt.$$

This expression vanishes in view of Proposition 6.1.C above. This completes the proof.

$\square$

6.2   Lagrangian intersections

We say that a Lagrangian submanifold $N \subset V$ has the *Lagrangian intersection property* if N intersects its image under any exact Lagrangian isotopy. In view of 6.1.A above one can reformulate this as $N \cap \phi(N) \neq \emptyset$ for all $\phi \in \mathrm{Ham}(V, \omega)$, or in other words the displacement energy of N is infinite: $e(N) = +\infty$.

Examples

6.2.A. Infinitesimal Lagrangian intersection problem
Let F be an autonomous Hamiltonian function on V and let $\xi = \mathrm{sgrad}F$ be its Hamiltonian vector field. Then ξ is tangent to N exactly at critical points of $F|_N$ (exercise). Since N is closed, $F|_N$ must have critical points and hence one cannot displace N by an infinitesimal Hamiltonian isotopy.

6.2.B. Gromov's Theorem
If $\pi_2(V, N) = 0$ and V has "nice" behaviour at infinity (say V is a product of a closed manifold and a cotangent bundle), Gromov [G1] (see also Floer

[F]) showed that N has the Lagrangian intersection property. In particular, this applies to the circle $\{r = 0\}$ in T^*S^1 (of course, this can be shown by an elementary area control, see Figure 7 and the discussion above). More generally, this holds for every non-contractible curve on an oriented surface. A proof of Gromov's theorem was sketched in 3.2.G above. We refer to [AL, Chap. X] for the details.

If $\pi_2(V, N) \neq 0$, the Lagrangian intersection property can be violated. Take for example a circle N of tiny area on $V = S^2$ and displace it, see Figure 8.

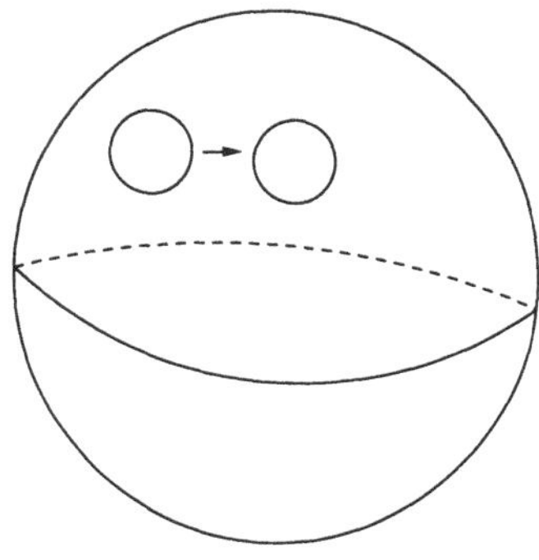

Figure 8

However, the Lagrangian intersection property obviously holds for the equator (use that the equator divides the sphere into two discs of equal area).

Definition 6.2.C. Let L be a closed Lagrangian submanifold of a symplectic manifold (M, Ω). We say that L has the *stable Lagrangian intersection property* if $L \times \{r = 0\}$ has the Lagrangian intersection property in $(M \times T^*S^1, \Omega + dr \wedge dt)$.

Let us give two examples which play an important role in the next chapters.

6.2.D. Tori in $T^*\mathbb{T}^n$.
Consider a Lagrangian torus in the cotangent bundle $T^*\mathbb{T}^n$ endowed with the standard symplectic structure (see 3.1.C above). Assume that it is homologous to the zero section. It is easy to check that the topological assumption of Theorem 6.2.B holds in this situation. Therefore such tori have the stable Lagrangian intersection property.

6.2.E The equator of S^2.
The stable Lagrangian intersection property also holds for equators of S^2. This follows from a difficult theorem due to Oh ([O1], [O2]) based on a clever version of Floer homology.

I do not know an example of a closed connected Lagrangian submanifold which has the Lagrangian intersection property, but does not have the stable one.

6.3   An application to Hamiltonian loops

Let (M, Ω) be a symplectic manifold. Assume that $L \subset M$ is a closed Lagrangian submanifold which has the Lagrangian intersection property. Let $\{g_t\}$ be a loop of Hamiltonian diffeomorphisms generated by a Hamiltonian $G \in \mathcal{H}$. Assume in addition that

- $g_t(L) = L$ for all $t \in S^1$;
- $G(x, t) = 0$ for all $x \in L, t \in S^1$.

An obvious example of such a loop is the constant loop $g_t \equiv \mathbb{1}$. A less straightforward example will be given in 6.3.C below.

Theorem 6.3.A. *Let $\{h_t\}$ be any other loop of Hamiltonian diffeomorphisms homotopic to $\{g_t\}$ as above. Let $H \in \mathcal{H}$ be its Hamiltonian function. Then there exist $x \in L$ and $t \in S^1$ such that $H(x, t) = 0$.*

As we will see in the next chapter, this result gives rise to non-trivial lower bounds for Hofer's distance.

Proof. Apply twice the Lagrangian suspension construction to L, first using $\{g_t\}$ and then using $\{h_t\}$. Denote by N_G and N_H the corresponding Lagrangian submanifolds of $M \times T^* S^1$. From the formula for the Lagrangian suspension we learn that $N_G = L \times \{r = 0\}$. Theorem 6.1.B above implies that N_H is isotopic to N_G by an exact Lagrangian isotopy. Thus the stable Lagrangian intersection property yields that $N_H \cap N_G \neq \emptyset$. Let $(x, 0, t)$, $x \in L$ be the intersection point. Since it lies on N_H we have that $x = h_t y$ and $0 = -H(h_t y, t, 0)$ for some $y \in L$. We conclude that $H(x, t) = 0$. $\qquad\square$

Recall that $\mathcal{H}_c$ denotes the space of all 1-periodic Hamiltonians which generate a contractible loop of Hamiltonian diffeomorphisms. As an immediate consequence of the theorem above we get the following result.

Corollary 6.3.B. *Let $L \subset M$ be a closed Lagrangian submanifold which has the stable Lagrangian intersection property. Then for every $H \in \mathcal{H}_c$ there exist $x \in L$ and $t \in S^1$ such that $H(x, t) = 0$.*

For instance this is true when $M = S^2$ and L is the equator of S^2. Note that the statement of the corollary in general becomes wrong if we do not assume the (stable) Lagrangian intersection property.

Example 6.3.C. Consider the Euclidean space $\mathbb{R}^3(x_1, x_2, x_3)$. Let $M = S^2$ be the unit sphere in the space endowed with the induced area form. The full

1-turn rotation around the x_3-axis is a Hamiltonian loop generated by the normalized Hamiltonian function $F_1(x) = 2\pi x_3$. (The reader can check this using our calculation in 1.4.H above.) Thus the k-turn rotation is generated by Hamiltonian $F_k(x) = 2\pi k x_3$. Since F_k vanishes identically on the equator $L = \{x_3 = 0\}$, it must vanish on *every* simple closed curve on S^2 which divides the sphere into two parts of equal areas. Note now that when k is even the k-turn rotation of S^2 represents a contractible loop in $SO(3)$, and therefore in $\text{Ham}(S^2)$. On the other hand $F_k \equiv 2\pi k\epsilon$ on the circle $C_\epsilon = \{x_3 = \epsilon\}$. We conclude that the phenomenon described in 6.3.B above is very rigid. It disappears completely when one considers circles which divide the sphere into parts of arbitrarily close but non-equal areas. Indeed, given any positive ϵ, choose k such that F_k restricted to C_ϵ is arbitrarily large! The crucial point, of course, is that the circle C_ϵ does not have the Lagrangian intersection property. One can displace it by an element of $SO(3)$ (cf. Figure 8 above).

Chapter 7

Diameter

In the present chapter we prove that the group of Hamiltonian diffeomorphisms of a closed oriented surface has infinite diameter with respect to Hofer's metric.

7.1   The starting estimate

Let (M, Ω) be a symplectic manifold, and let $L \subset M$ be a closed Lagrangian submanifold with the stable Lagrangian intersection property. Let $F \subset \mathcal{F}$ be a Hamiltonian function such that $|F(x,t)| \geq C$ for all $x \in L$ and $t \in S^1$. Here C is a positive constant. The next proposition gives a lower bound for the quantity $l(F) = \tilde{\rho}(\mathbb{1}, \tilde{\phi}_F)$ introduced in 5.3 above.

Proposition 7.1.A. *Under these assumptions $l(F) \geq C$.*

Proof. This is an immediate consequence of 5.3.A and 6.3.B above. Indeed, 6.3.B states that every function $H \in \mathcal{H}_c$ vanishes at some point (y, τ) where $y \in L$ and $\tau \in S^1$. Thus $|F(y, \tau) - H(y, \tau)| \geq C$ and hence $|||F - H||| \geq C$. Since this holds true for every $H \in \mathcal{H}_c$ we conclude from 5.3.A that $l(F) \geq C$. $\square$

We wish to extend the estimate above to $\rho(\mathbb{1}, \phi_F)$. Note that if the group $\mathrm{Ham}(M, \Omega)$ is simply connected with respect to the (strong Whitney) C^∞-topology then all paths with common end points are homotopic and therefore $l(F) = \rho(\mathbb{1}, \phi_F)$. If the fundamental group $\pi_1(\mathrm{Ham}(M, \Omega))$ is non-trivial then in general there is no hope to extend estimate 7.1.A without some additional information. Indeed, it could happen that in the situation described above there exists a shorter path joining $\mathbb{1}$ with ϕ_F, which of course cannot be homotopic to the flow $\{f_t\}$, $t \in [0; 1]$ generated by F. It turns out however that in some interesting cases one can go round this difficulty. In order to do this we have to look more closely at the fundamental group of $\mathrm{Ham}(M, \Omega)$.

7.2 The fundamental group

Not much is known about $\pi_1(\mathrm{Ham}(M,\Omega))$. There is a complete picture for surfaces (based on classical methods) and some four-dimensional manifolds [G1], [A], [AM] (based on Gromov's theory of pseudo-holomorphic curves). In higher dimensions, only a few partial results are available (actually, I do not know any symplectic manifold M of dimension ≥ 6 for which one can completely describe the fundamental group of $\mathrm{Ham}(M,\Omega)$). For instance, it is known that $\mathrm{Ham}(\mathbb{R}^{2n})$ is simply connected (and in fact, contractible) for $n = 1, 2$, while nothing is known already for $n = 3$. We will need the information about $\pi_1(\mathrm{Ham}(M,\Omega))$, where M is a closed orientable surface.

7.2.A. The sphere (cf. 1.4.H and 6.3.C above).
The inclusion $SO(3) \to \mathrm{Ham}(S^2)$ induces the isomorphism of fundamental groups. In particular, $\pi_1(\mathrm{Ham}(S^2))$ equals $\mathbb{Z}_2$. The non-trivial element is generated by the 1-turn rotation of the sphere around the vertical axis.

7.2.B. Surfaces of genus ≥ 1.
In this case one can show that the group of Hamiltonian diffeomorphisms is simply connected.

These facts are well known to experts, but it seems to me that they have not appeared in published form. Thus I will outline the argument and give some auxiliary references with the hope that this will enable the reader to reconstruct the proof. We write $\mathrm{Diff}_0(M)$ (resp. $\mathrm{Symp}_0(M)$) for the identity component of the group of all (resp. symplectic) diffeomorphisms of surface M.

Sketch of the proof:

1) *The inclusion $\pi_1(\mathrm{Symp}_0(M)) \to \pi_1(\mathrm{Diff}_0(M))$ is an isomorphism.* To see this, consider the space X of all area forms on M with total area 1. Fix an area form $\Omega \in X$. Consider the mapping $\mathrm{Diff}_0(M) \to X$ which sends a diffeomorphism f to the form $f^*\Omega$. One can adjust the proof of Moser's stability theorem [MS, pp. 94–95] in order to show that this mapping is a Serre fibration. Note that its fiber is nothing else but $\mathrm{Symp}_0(M,\Omega)$, and the base X is contractible. The desired fact now follows from the homotopy exact sequence of the fibration.

2) *The topology of $\mathrm{Diff}_0(M)$ is known* (see [EE]). In particular, this group is contractible for surfaces of genus ≥ 2. Further, when $M = S^2$ it has $SO(3)$ as strong deformation retract, and when $M = \mathbb{T}^2$ it has $\mathbb{T}^2$ as strong deformation retract (here the torus acts on itself by translations).

3) *The inclusion $j : \pi_1(Ham(M,\Omega)) \to \pi_1(\mathrm{Symp}_0(M,\Omega))$ is injective* (see [MS], 10.18 (iii)). In fact this is true for all closed symplectic manifolds.

4) Let us combine these facts. We immediately get 7.2.B for surfaces of genus ≥ 2. Taking into account that $\mathrm{Symp}_0(S^2) = \mathrm{Ham}(S^2)$ (see 1.4.C) yields 7.2.A. It remains to handle the case of the torus $\mathbb{T}^2$.

5) Fix a point $y \in \mathbb{T}^2$, and consider the evaluation map $\mathrm{Diff}_0(\mathbb{T}^2) \to \mathbb{T}^2$ which sends a diffeomorphism f to $f(y)$. It induces a map $e_D : \pi_1(\mathrm{Diff}_0(\mathbb{T}^2)) \to \pi_1(\mathbb{T}^2)$. It is easy to see from step 2 that e_D is an isomorphism. Consider now the restrictions of the evaluation map to $\mathrm{Ham}(\mathbb{T}^2)$ and $\mathrm{Symp}_0(\mathbb{T}^2)$, and denote by e_H and e_S respectively the induced homomorphisms of the fundamental groups. Step 1 implies that e_S is an isomorphism. Using step 3 we get that $e_H = e_S \circ j$, where j is injective. It follows from a theorem by Floer that e_H vanishes (see [LMP1]). Thus $\pi_1(\mathrm{Ham}(\mathbb{T}^2)) = 0$. The proof is complete. $\qquad\square$

Theorem 7.2.C. *([P5]) Assume that*

- *either $M = S^2$ and $L \subset S^2$ is an equator,*

- *or M is a closed orientable surface of genus ≥ 1 and L is a non-contractible closed curve.*

Let $F \in \mathcal{F}$ be a function such that $|F(x,t)| \geq C$ for all $x \in L$ and $t \in S^1$, where C is a positive constant. Then $\rho(\mathbb{1}, \phi_F) \geq C$.

Proof. If M has genus ≥ 1 then $l(F) = \rho(\mathbb{1}, \phi_F)$ since the group $\mathrm{Ham}(M, \Omega)$ is simply connected (use 7.2.B). Thus the result follows from 7.1.A above. If $M = S^2$ then the non-trivial element of $\pi_1(\mathrm{Ham}(M, \Omega))$ is represented by the 1-turn rotation (see 7.2.A). Its Hamiltonian vanishes on L (see 6.3.C). Thus Theorem 6.3.A implies that every function from $\mathcal{H}$ must vanish at some point of $L \times S^1$. The needed statement now follows from 5.1.B above. $\qquad\square$

Corollary 7.2.D. *The group of Hamiltonian diffeomorphisms of a closed surface has infinite diameter with respect to Hofer's metric.*

Indeed, let M and L be as in 7.2.C above, and let $B \subset M$ be an open disc disjoint from L. Take a time-independent Hamiltonian $F \in \mathcal{F}$ which is identically equal to C outside B. The theorem above implies that $\rho(\mathbb{1}, \phi_F) \geq C$. Taking C arbitrarily large we get that the diameter is infinite. Note also that in this example the support of ϕ_F is contained in B. Thus shrinking B and simultaneously increasing C we get a sequence of Hamiltonian diffeomorphisms which converges to the identity in the C^0-sense but diverges in Hofer's metric.

Let me mention that for a closed surface M of genus ≥ 1 there exist at least two other different proofs of the fact that the diameter of $\mathrm{Ham}(M, \Omega)$ is infinite. One of them goes as follows.

Exercise 7.2.E. (see [LM2]). Let F be a normalized Hamiltonian on M. Assume that some regular level set of F contains a non-contractible closed curve.

Consider the lift $\tilde{f}_t$ of the corresponding Hamiltonian flow f_t to the universal cover $\tilde{M}$ of M. Show that there exists $\epsilon > 0$ and a family of discs $D_t \subset \tilde{M}$ of area ϵt such that $\tilde{f}_t D_t \cap D_t = \emptyset$ for all sufficiently large t. Conclude from Theorem 3.2.C above that $\rho(\mathbb{1}, f_t) \to +\infty$ when $t \to +\infty$.

Another proof (see [Sch3]) is based on Floer homology (see Chapter 13 for applications of Floer homology to Hofer's geometry).

Finally, the fact that diam $\mathrm{Ham}(M, \Omega) = +\infty$ has been proved for some higher-dimensional manifolds (see [LM2], [Sch3], [P5]). In general, however, this is still an open problem.

7.3   The length spectrum

As we have seen above, our method provides a lower bound for $\rho(\mathbb{1}, \phi_F)$ when we have very precise information on the fundamental group of $\mathrm{Ham}(M, \Omega)$. It was already mentioned that no such information is available in dimension ≥ 6. Now we will slightly modify the way of arguing and extend the class of manifolds where the method works. The next notion is one of the main characters of our story.

Definition 7.3.A. For an element $\gamma \in \pi_1(\mathrm{Ham}(M, \Omega))$ define the *norm* of γ by

$$\nu(\gamma) = \inf \mathrm{length}\{h_t\},$$

where the infimum is taken over all Hamiltonian loops $\{h_t\}$ which represent γ. The set

$$\{\nu(\gamma) | \gamma \in \pi_1(\mathrm{Ham}(M, \Omega))\}$$

is called the *length spectrum* of $\mathrm{Ham}(M, \Omega)$.

Exercise
  (i) Show that $\pi_1(\mathrm{Ham}(M, \Omega))$ is always an abelian group (use the same argument as for finite-dimensional Lie groups). Thus we denote the group operation by $+$, and write 0 for the neutral element.

  (ii) Prove that $\nu(\gamma) = \nu(-\gamma)$, and $\nu(\gamma + \gamma') \leq \nu(\gamma) + \nu(\gamma')$.

No general statement about the non-degeneracy of ν is known at the moment (and I would not be surprised by an example of $\gamma \neq 0$ with $\nu(\gamma) = 0$, cf. example 7.3.B below). Thus, frankly speaking, ν is a pseudo-norm, though we call it norm. In Chapter 9 below we will describe a method which provides lower bounds for $\nu(\gamma)$, and in particular enables us to calculate the length spectrum for S^2.

Example 7.3.B. Liouville manifolds. We say that an open symplectic manifold (M, Ω) has the *Liouville property* if there exists a smooth family of diffeomorphisms

$$D_c : M \to M, \ c \in (0; +\infty)$$

such that $D_1 = \mathbb{1}$ and $D_c^* \Omega = c\Omega$ for all c. Such diffeomorphisms D_c are of course not compactly supported for $c \neq 1$. An important class of examples is given by cotangent bundles endowed with the standard symplectic structure (see 3.1.C above). The diffeomorphism D_c in this situation is just the fiberwise homothety $(p, q) \to (cp, q)$. We claim that the length spectrum of $\mathrm{Ham}(M, \Omega)$ equals $\{0\}$ provided (M, Ω) has the Liouville property. The proof of the claim is based on the following simple fact.

Exercise. Let $\{f_t\}$ be the Hamiltonian flow generated by a normalized Hamiltonian $F(x, t)$. Then for every $c > 0$ the flow $\{D_c f_t D_c^{-1}\}$ is again a Hamiltonian flow whose normalized Hamiltonian equals $cF(D_c^{-1}x, t)$.

Let $\{h_t\}$ be a loop of Hamiltonian diffeomorphisms. Consider the family of homotopic loops $\{D_c h_t D_c^{-1}\}$. It follows from the exercise above that the lengths of the loops go to zero when $c \to 0$, thus every loop can be homotoped to a loop of an arbitrary small length. We conclude that the length spectrum is $\{0\}$ (without any knowledge about the fundamental group). In the next chapter we will discuss this example in the context of classical mechanics.

7.4 Refining the estimate

Theorem 7.4.A. *Let (M, Ω) be a symplectic manifold and let $L \subset M$ be a closed Lagrangian submanifold with the stable Lagrangian intersection property. Assume that the length spectrum of $\mathrm{Ham}(M, \Omega)$ is bounded from above by some $K \geq 0$. Let $F \in \mathcal{F}$ be a function such that $|F(x, t)| \geq C$ for all $x \in L$ and $t \in S^1$. Then*

$$\rho(\mathbb{1}, \phi_F) \geq C - K.$$

Proof. Fix an arbitrary $\epsilon > 0$. Let $\{g_t\}$ be a path of Hamiltonian diffeomorphisms which joins $\mathbb{1}$ with ϕ_F. Consider the loop $\{f_t \circ g_t^{-1}\}$. By our assumption this loop is homotopic to a loop $\{h_t\}$ whose length does not exceed $K + \epsilon$. The path $\{f_t\}$ is homotopic with fixed end points to the composition $\{h_t \circ g_t\}$. Therefore

$$l(F) \leq \mathrm{length}\{h_t\} + \mathrm{length}\{g_t\}.$$

Since $l(F) \geq C$ in view of 7.1.A we conclude that $\mathrm{length}\{g_t\} \geq C - K - \epsilon$. Thus $\rho(\mathbb{1}, \phi_F) \geq C - K$. $\qquad\square$

Chapter 8

Growth and Dynamics

In this chapter we discuss the asymptotic geometric behaviour of one-parameter subgroups of the group of Hamiltonian diffeomorphisms. We describe a link between geometry and invariant tori of classical mechanics.

8.1 Invariant tori of classical mechanics

Invariant Lagrangian tori of Hamiltonian dynamical systems play an important role in classical mechanics. We start with an obstruction to the existence of invariant tori which comes from the geometry of the group of Hamiltonian diffeomorphisms (see 8.1.C below).

Consider the n-dimensional torus $\mathbb{T}^n$ endowed with the Euclidean metric $ds^2 = \Sigma_{j=1}^n dq_j^2$. The Euclidean geodesic flow can be described by a Hamiltonian system on the cotangent bundle $T^*\mathbb{T}^n$ endowed with the standard symplectic form $\Omega = dp \wedge dq$. The Hamiltonian function is given by $F(p,q) = \frac{1}{2}|p|^2$. Solving the Hamiltonian system

$$\begin{cases} \dot{p} = 0 \\ \dot{q} = p \end{cases}$$

we find the Hamiltonian flow

$$f_t(p,q) = (p, q + pt).$$

Let us describe the dynamics of this flow. Every torus $\{p = a\}$ is invariant under $\{f_t\}$, and moreover the restriction of the flow to each such torus is simply the (quasi)-periodic motion $q \to q + at$. All these tori are homologous to the zero section of the cotangent bundle. Geometrically they correspond to families of parallel Euclidean lines on $\mathbb{T}^n$.

The flow above belongs to an important class of time-independent Hamiltonian systems called *integrable* systems (see [Ar]). Integrable systems can be characterized by the fact that their energy levels are foliated (up to measure zero) by invariant middle-dimensional tori which carry (quasi)-periodic motion. Traditionally this is considered as the simplest pattern of dynamical behaviour in classical mechanics. A natural question is what happens to the invariant tori when one perturbs the system. The Kolmogorov-Arnold-Moser theory (or in short KAM theory, see [Ar]) tells us that for small perturbations most of the invariant tori persist. One has to require, however, that the rotation vector is "sufficiently irrational". This means, in particular, that if in some angular coordinates θ on an invariant torus the dynamics is given by $\dot{\theta} = a$, where $a = (a_1, ..., a_n) \in \mathbb{R}^n$, then $a_1, ..., a_n$ should be linearly independent over $\mathbb{Q}$. For large perturbations the "topologically essential" tori may disappear. For instance, one can deform the Euclidean metric on $\mathbb{T}^n$ to a Riemannian metric such that its geodesic flow has no invariant tori which carry quasi-periodic motion and are homologous to the zero section (see [AL]).

Note that in our starting example the invariant tori $\{p = a\}$ are Lagrangian. This is a general phenomenon which we are going to explain now.

Exercise 8.1.A. Let $F : M \to \mathbb{R}$ be a time independent Hamiltonian on a symplectic manifold M. Consider a closed submanifold $L \subset \{H = c\}$. Show that if L is Lagrangian then L is invariant under the Hamiltonian flow of F. *Hint:* use linear algebra in order to prove that sgrad F must be tangent to L.

In some interesting cases the statement above can be reversed.

Proposition 8.1.B. *([He]) Let $F : T^*\mathbb{T}^n \to \mathbb{R}$ be a time independent Hamiltonian. Consider an invariant torus $L \subset \{H = c\}$ carrying a quasi-periodic motion $\dot{\theta} = a$ where the coordinates of a are, as before, linearly independent over $\mathbb{Q}$. Then L is Lagrangian.*

Proof. Take a point $x \in L$, and assume that $\Omega|_{T_x L}$ has the form $\sum b_{ij} d\theta_i \wedge d\theta_j$. Since the dynamics is just a shift, $\theta(t) = \theta(0) + at$, we get that $\Omega|_{T_y L} = \sum b_{ij} d\theta_i \wedge d\theta_j$ for every point y which lies on the trajectory of x. Note that every trajectory is dense on the torus, so we conclude that $\Omega = \sum b_{ij} d\theta_i \wedge d\theta_j$ *everywhere on L.* But $\Omega|_{TL}$ is an exact two-form. Hence $b_{ij} = 0$ for all i, j. Therefore $\Omega|_{TL} = 0$ and we have shown that L is Lagrangian. $\qquad\square$

Let F be a time-independent compactly supported Hamiltonian function on $(T^*\mathbb{T}^n, \Omega)$. Define a number $E(F) = \sup |E|$, where the supremum is taken over all E such that the energy level $\{F = E\}$ *contains a Lagrangian torus homologous to the zero section.* How can one find a non-trivial upper bound for $E(F)$? Questions of this flavor are studied in the framework of converse KAM theory.[1] Write $\{f_t\}$ for the Hamiltonian flow generated by F.

Theorem 8.1.C. *(cf. [BP2],[P8])* $\rho(\mathbb{1}, f_t) \geq E(F)t$ *for all* $t \in \mathbb{R}$.

Proof. Note that it suffices to prove the inequality for $t = 1$ (make a time reparametrization). Let $L \subset \{H = E\}$ be a Lagrangian torus homologous to the zero section. Then L has the stable Lagrangian intersections property (see 6.2.D). Since $(T^*\mathbb{T}^n, \Omega)$ is a Liouville manifold the length spectrum of $\mathrm{Ham}(M, \Omega)$ equals $\{0\}$ (see 7.3.B). Therefore all the assumptions of Theorem 7.4.A are satisfied. This theorem implies that $\rho(\mathbb{1}, f_1) \geq E$. Taking the supremum over all such E, we get the required estimate. $\qquad\qquad\square$

The geometric content of this estimate should be understood in a more general context of the growth of one-parameter subgroups of Hamiltonian diffeomorphisms.

8.2 Growth of one-parameter subgroups

Let (M, Ω) be a symplectic manifold, and let $\{f_t\}$ be a one-parameter subgroup of $\mathrm{Ham}(M, \Omega)$ generated by a normalized Hamiltonian function $F \in \mathcal{A}$. One of the central problems of Hofer's geometry is to explore interrelations between the function $\rho(\mathbb{1}, f_t)$ and the dynamics of the flow $\{f_t\}$. For instance, Theorem 8.1.C above states that invariant tori of an autonomous Hamiltonian flow on $T^*\mathbb{T}^n$ which are homologous to the zero section and carry quasi-periodic motion contribute to the linear growth of the function $\rho(\mathbb{1}, f_t)$. There exists another, purely geometrical reason for the interest in this function. It comes from the theory of geodesics of Hofer's metric (see Chapter 12 below). A Hamiltonian path $\{f_t\}$ is called *a strictly minimal geodesic* if each of its segments minimizes the length between its end points. Conjecturally (see 12.6.A below) all *sufficiently short* segments of any one-parameter subgroup are strictly minimal geodesics (in other words, every one-parameter subgroup is locally strictly minimal). However as we will see in 8.2.H *long segments* may lose the minimality. The minimality breaking is an intriguing phenomenon which is still far from being understood. At present several approaches to this phenomenon are known. One of them, based on the theory of conjugate points in Hofer's geometry is presented in Chapter 12. Here we discuss an approach based on the notion of *asymptotic growth* of a one-parameter subgroup (see [BP2]). The asymptotic growth is defined as

$$\mu(F) = \lim_{t \to +\infty} \frac{\rho(\mathbb{1}, f_t)}{t\|F\|}.$$

---

[1] The goal of KAM theory is to prove existence of invariant Lagrangian tori, while converse KAM studies obstructions to their existence, see e.g. [Mac] and references therein.

Exercise. Show that the limit above exists. *Hint:* Use that the function $\rho(\mathbb{1}, f_t)$ is subadditive: $\rho(\mathbb{1}, f_{t+s}) \leq \rho(\mathbb{1}, f_t) + \rho(\mathbb{1}, f_s)$.

Clearly, $\mu(F)$ always belongs to $[0; 1]$. If $\mu(F) < 1$ then the path $\{f_t\}$ is not a strictly minimal geodesic.

Let us consider several examples of the behaviour of the function $\rho(\mathbb{1}, f_t)$. It was proved by Hofer [H2] that every one-parameter subgroup of $\mathrm{Ham}(\mathbb{R}^{2n})$ is locally strictly minimal. On the other hand Sikorav [S2] discovered the striking fact that each such subgroup remains a bounded distance of identity, and thus cannot be globally strictly minimal.

Theorem 8.2.A. *Let $\{f_t\}$ be a one-parameter subgroup of* $\mathrm{Ham}(\mathbb{R}^{2n})$ *generated by a compactly supported Hamiltonian function F. Assume that the support of F is contained in a Euclidean ball of radius r. Then the function $\rho(\mathbb{1}, f_t)$ is bounded:* $\rho(\mathbb{1}, f_t) \leq 16\pi r^2$.

We refer the reader to [HZ, p. 177] for the detailed proof of this theorem (see also discussion in 12.6.E below).

Let us return now to one-parameter subgroups of $\mathrm{Ham}(T^*\mathbb{T}^n)$. First of all, it is proved in [LM2] that all of them are locally strictly minimal. In other words $\rho(\mathbb{1}, f_t) = t\|F\|$ provided t is small enough. Of course, this implies that our estimate 8.1.C above is not sharp for small t. Indeed, in general $E(F)$ is strictly smaller than $\|F\|$. Nevertheless, in the case when $n = 1$ and $F \geq 0$ estimate 8.1.C is asymptotically sharp! Notice that since every closed curve on the cylinder $T^*\mathbb{T}^1$ is Lagrangian, the quantity $E(F)$ is simply the supremum of those real numbers E for which the level $\{F = E\}$ contains a non-contractible embedded circle.

Theorem 8.2.B. *([PS]) Let F be a non-negative compactly supported Hamiltonian on the cylinder $T^*\mathbb{T}^1$ with $\|F\| = 1$. Then the converse KAM parameter $E(F)$ coincides with the asymptotic growth $\mu(F)$:*

$$E(F) = \mu(F).$$

Proof. We have to prove that $\mu(F) \leq E(F)$. Combining this with 8.1.C we get the desired result. If $E(F) = \max F = 1$ then 8.1.C yields that $\mu(F) = E(F)$. Suppose now that $E(F) < 1$. The idea is to decompose the flow $\{f_t\}$ into a product of two commuting flows with simple asymptotic behaviour. Choose $\epsilon > 0$ small enough, and consider a smooth non-decreasing function $u : [0; +\infty) \to [0; +\infty)$ with the following properties:

- $u(s) = s$ for $s \leq E(F) + \epsilon$;
- $u(s) = E(F) + 2\epsilon$ for $s \geq E(F) + 3\epsilon$;
- $u(s) \leq s$ for all s.

Consider the new Hamiltonians $G = u \circ F$ and $H = F - G$, and denote by $\{g_t\}$ and $\{h_t\}$ the corresponding Hamiltonian flows. These flows commute and $f_t = g_t h_t$. Thus

$$\rho(\mathbb{1}, f_t) \leq \rho(\mathbb{1}, g_t) + \rho(\mathbb{1}, h_t). \qquad (8.2.\text{C})$$

Note that $\|G\| \leq E(F) + 2\epsilon$, thus

$$\rho(\mathbb{1}, g_t) \leq t(E(F) + 2\epsilon). \qquad (8.2.\text{D})$$

Further, the support of H is contained in a subset $D_\epsilon = \{F \geq E(F) + \epsilon\}$. For a generic sufficiently small ϵ the set D_ϵ is a domain whose boundary consists of *contractible* closed curves (we use here the definition of $E(F)$). Assume now that the support of F is contained in an annulus

$$A = \{(p, q) \in T^*\mathbb{T}^1 \mid |q| \leq a/2\}$$

for some $a > 0$. Note that $\partial D_\epsilon \subset A$. Hence the set D_ϵ is contained in some set $D' \subset A$ which is a finite union of pairwise disjoint closed discs of total area at most a. Since the cylinder has the infinite area, it is an easy consequence of the Dacorogna-Moser theorem [HZ, Sect. 1.6] that there exists a symplectic embedding $i : \mathbb{R}^2 \to T^*\mathbb{T}^1$, and a finite union $D'' \subset \mathbb{R}^2$ of Euclidean discs such that i maps D'' diffeomorphically on D'. Clearly, i induces the natural homomorphism

$$i_* : \mathrm{Ham}(\mathbb{R}^2) \to \mathrm{Ham}(T^*\mathbb{T}^1).$$

It is important to notice that i_* does not increase the corresponding Hofer distances. Our flow h_t lies in the image of i_*, i.e., $h_t = i_*(e_t)$ where e_t is a one–parameter subgroup of $\mathrm{Ham}(\mathbb{R}^2)$ whose Hamiltonian is supported in D''. Thus Theorem 8.2.A above implies that

$$\rho(\mathbb{1}, h_t) \leq 16a.$$

Combining this inequality with (8.2.D) and (8.2.C) we get that

$$\rho(\mathbb{1}, f_t) \leq t(E(F) + 2\epsilon) + 16a \qquad (8.2.\text{E})$$

for all $t > 0$. Dividing by t and passing to the limit when $t \to +\infty$ we get that $\mu(F) \leq E(F) + 2\epsilon$. Since ϵ is arbitrary small, this completes the proof. $\square$

Remark 8.2.F. The same proof shows that if $E(F) = 0$ then the function $\rho(\mathbb{1}, f_t)$ is bounded. Indeed, since (8.2.E) holds for all $\epsilon > 0$ we get that $\rho(\mathbb{1}, f_t) \leq 16a$. In view of the inequality $E(F) \leq \mu(F)$ this implies the following "rigidity"-type statement: *if $\mu(F) = 0$ then $\rho(\mathbb{1}, f_t)$ is bounded* (see 8.4 below for further discussion).

Theorem 8.2.B and Remark 8.2.F hold true for all open surfaces of infinite area (see [PS]). Moreover, one can easily modify the definition of $E(F)$ and extend these statements to arbitrary (not necessarily non-negative) Hamiltonians F. At the moment, no generalization of 8.2.B to higher dimensions is known. The next discussion shows however that estimate 8.1.C is sharp at least in the following very special situation.

Let $F : T^*\mathbb{T}^n \to \mathbb{R}$ be a compactly supported time independent Hamiltonian function which satisfies the following conditions:

(i) $F \geq 0$

(ii) $\max F = 1$

(iii) the maximum set $\Sigma = \{F = 1\}$ is a smooth section of the cotangent bundle.

It turns out that the geometry of the corresponding flow $\{f_t\}$ drastically depends on whether Σ is Lagrangian or not!

Suppose that Σ is a Lagrangian submanifold. Then by definition $E(F) = 1$. Therefore Theorem 8.1.C above implies that $\{f_t\}$ is a strictly minimal geodesic, and in particular $E(F) = \mu(F)$.

Assume now that Σ is not Lagrangian, that is Ω does not vanish at least on one tangent space to Σ. In this case it is unknown whether $\mu(F)$ equals $E(F)$ or not. However we claim that estimate 8.1.C is at least *non-trivial*, namely

$$E(F) \leq \mu(F) < 1. \tag{8.2.G}$$

In order to explain this inequality, we need the following fairly general result which in some interesting situations allows us to show that $\mu(F)$ is strictly less than 1.

Theorem 8.2.H. *Let F be a time-independent normalized Hamiltonian on a symplectic manifold (M, Ω). Let Σ_+ and Σ_- be the maximum and the minimum sets of F respectively. Suppose that there exists $\phi \in \mathrm{Ham}(M, \Omega)$ such that either $\phi(\Sigma_+) \cap \Sigma_+ = \emptyset$ or $\phi(\Sigma_-) \cap \Sigma_- = \emptyset$. Then $\mu(F) < 1$, and in particular the Hamiltonian flow of F is not a strictly minimal geodesic.*

The proof is given in 8.3 below.

Let us sketch the proof of 8.2.G. One can show that since Σ is not Lagrangian there exists a Hamiltonian diffeomorphism ϕ such that $\phi(\Sigma) \cap \Sigma = \emptyset$. Thus combining Theorems 8.1.C and 8.2.H we get that $E(F) \leq \mu(F) < 1$. The proof of the existence of the Hamiltonian diffeomorphism ϕ which displaces the non-Lagrangian submanifold Σ is complicated. It is based on Gromov's h-principles for partial differential relations. In fact one can even show more, namely that the displacement energy of Σ vanishes. We refer the reader to [P2], [LS] for the proof of this result and its generalizations.

Let us mention also that Siburg [Si2] extended inequality 8.1.C to *time-dependent* Hamiltonian flows. Note that in the time-dependent case it is already non-trivial to define the converse KAM-type parameter because of the lack of the energy conservation. Siburg's definition is based on the concept of minimal action developed by Mather.

8.3 Curve shortening in Hofer's geometry

Let (M, Ω) be a symplectic manifold. For every function $F \in \mathcal{A}$, $F \neq 0$ set

$$\delta(F) = \inf_{\phi} \frac{\|F + F \circ \phi\|}{2\|F\|},$$

where the infimum is taken over all $\phi \in \mathrm{Ham}(M, \Omega)$.

Theorem 8.3.A. *([BP2])*
$$\mu(F) \leq \delta(F).$$

Theorem 8.2.H above is an immediate consequence of this result. Indeed, if F satisfies the assumptions of 8.2.H then $\delta(F) < 1$.

Proof of 8.3.A. Without loss of generality, assume that $\|F\| = 1$. Take $\phi \in \mathrm{Ham}(M, \Omega)$ and $T > 0$. Write

$$f_{2T} = \left(f_T \circ \phi \circ f_T \circ \phi^{-1} \right) \circ \left(\phi \circ f_T^{-1} \circ \phi^{-1} \circ f_T \right) = A \circ B.$$

Observe that B is a commutator so $\rho(\mathbb{1}, B) \leq 2\rho(\mathbb{1}, \phi)$. The diffeomorphism A is generated by a path $g_t = f_t \phi f_t \phi^{-1}$ for $t \in [0, T]$ and the corresponding Hamiltonian is

$$G(x, t) = F(x) + F(\phi^{-1} f_t^{-1} x).$$

So we have

$$\begin{aligned}
\|G_t\| &= \|F + F \circ \phi^{-1} \circ f_t^{-1}\| \\
&= \|F \circ f_t + F \circ \phi^{-1}\| \\
&= \|F + F \circ \phi^{-1}\|
\end{aligned}$$

since $F \circ f_t = F$ (energy conservation). So

$$\rho(\mathbb{1}, f_{2T}) \leq T\|F + F \circ \phi^{-1}\| + 2\rho(\mathbb{1}, \phi)$$

and hence

$$\frac{\rho(\mathbb{1}, f_{2T})}{2T} \leq \frac{1}{2}\|F + F \circ \phi^{-1}\| + \frac{\rho(\mathbb{1}, \phi)}{T}$$

for all $\phi \in \mathrm{Ham}(M, \Omega)$. Now let $T \to \infty$ and deduce that $\mu(F) \leq \delta(F)$. $\qquad\square$

We refer the reader to [LM2] and [P9] for various curve shortening procedures in Hofer's geometry. We will return to this issue in Chapter 11 below. The quantity $\delta(F)$ admits the following natural generalization. Set

$$\delta_N(F) = \inf_{\phi_1,\ldots,\phi_{N-1}} \frac{1}{N} \frac{\|\sum_{j=0}^{N-1} F \circ \phi_j\|}{\|F\|},$$

where $\phi_0 = \mathbb{1}$ and the infimum is taken over all sequences $\{\phi_j\}$, $j = 1, \ldots, N-1$ of Hamiltonian diffeomorphisms. With this language $\delta = \delta_2$. One can show [P9] that on a closed symplectic manifold $\delta_N(F) \to 0$ when $N \to +\infty$ for every function $F \in \mathcal{A}$. I do not know the decay rate of the sequence $\delta_N(F)$.

8.4 What happens when the asymptotic growth vanishes?

As we have seen in 8.2.F above, on the cylinder (and more generally, on open surfaces of infinite area) every one-parameter subgroup whose asymptotic growth vanishes remains a bounded distance of the identity. What happens on other symplectic manifolds?

Problem 8.4.A. Does there exist a symplectic manifold (M, Ω) and a one-parameter subgroup $\{f_t\}$ of $\mathrm{Ham}(M, \Omega)$ such that the function $\rho(\mathbb{1}, f_t)$ has an intermediate asymptotic at infinity (for instance, it grows like $\sqrt{t}$)?

This problem is open even for such an elementary symplectic manifold as the 2-torus $\mathbb{T}^2$. The only excuse is that the functions $F \in \mathcal{A}(\mathbb{T}^2)$ with $\mu(F) = 0$ are *non-generic*, as the next result shows.

Theorem 8.4.B. *Assume that 0 is a regular value of $F \in \mathcal{A}(\mathbb{T}^2)$. Then $\mu(F) > 0$.*

Proof. The set $D = \{F = 0\}$ consists of a finite number of pairwise disjoint embedded circles. Thus there exists a non-contractible simple closed curve $L \subset \mathbb{T}^2$ such that $L \cap D = \emptyset$. Thus for some $C > 0$ we have that $|F(x)| > C$ for all $x \in L$. Since $\pi_1(\mathrm{Ham}(\mathbb{T}^2)) = 0$ (see 7.2.B), it follows from 7.4.A that $\rho(\mathbb{1}, f_t) \geq Ct$ for all t, and in particular $\mu(F) > 0$. $\qquad\square$

Chapter 9

Length Spectrum

In this chapter we describe a method of calculation of the length spectrum in Hofer's geometry. Our approach is based on the theory of symplectic fibrations over the 2-sphere.

9.1 The positive and negative parts of Hofer's norm

Let (M, Ω) be a closed symplectic manifold. For $\gamma \in \pi_1(\mathrm{Ham}(M, \Omega))$ we set

$$\nu_+(\gamma) = \inf_F \int_0^1 \max_x F(x,t)dt = \inf_F \max_{x,t} F(x,t),$$

$$\nu_-(\gamma) = \inf_F \int_0^1 -\min_x F(x,t)dt = \inf_F \left(-\min_{x,t} F(x,t)\right),$$

where the infimum is taken over all normalized periodic Hamiltonians $F \in \mathcal{H}$ which generate a loop representing γ. The correctness of this definition can be proved exactly in the same way as Lemma 5.1.C above.

Exercise. Prove that $\nu_+(\gamma) = \nu_-(-\gamma)$, and $\nu(\gamma) \geq \nu_-(\gamma) + \nu_+(\gamma)$ (cf. the open problem in 2.4 above).

The calculation of $\nu_+(\gamma)$ turns out to be non-trivial even in the following simplest case. Consider (S^2, Ω) normalized so that $\int_{S^2} \Omega = 1$. Let γ be the class of the 1-turn rotation $\{f_t\}$, and let $F \in \mathcal{H}$ be the Hamiltonian generating $\{f_t\}$. We have seen in 6.3.C above that $\max F = -\min F = \frac{1}{2}$. (The factor 2π in 6.3.C disappeared in view of the normalization of Ω above.) Thus $\nu_+(\gamma) \leq \frac{1}{2}$. In fact equality holds!

Theorem 9.1.A. $\nu_+(\gamma) = \frac{1}{2}$.

A Hamiltonian loop $\{f_t\}$ representing a class $\gamma \neq 0$ is called a *closed minimal geodesic* if length$\{f_t\} = \nu(\gamma)$. Note that a closed minimal geodesic is never strictly minimal.

Corollary. *([LM2]) $\nu(\gamma) = 1$ and $\{f_t\}$ is a closed minimal geodesic.*

Proof. First of all, length$\{f_t\} = \frac{1}{2} - (-\frac{1}{2}) = 1$ and hence $\nu(\gamma) \leq 1$. Since $2\gamma = 0$ we deduce that

$$\nu_-(\gamma) = \nu_+(-\gamma) = \nu_+(\gamma) = \frac{1}{2}.$$

Furthermore, $\nu(\gamma) \geq \nu_-(\gamma) + \nu_+(\gamma) = 1$. We conclude that $\nu(\gamma) = 1$ and $\{f_t\}$ is a minimal geodesic. $\qquad\qquad\square$

Theorem 9.1.A is proved in 9.4 below. We refer to [P3] for generalizations to $\mathbb{C}P^n$, $n \geq 2$.

9.2   Symplectic fibrations over S^2

Let (M, Ω) be a closed symplectic manifold. We will assume from now on[1] that $H^1(M, \mathbb{R}) = 0$. As a consequence $\mathrm{Ham}(M, \Omega)$ coincides with the connected component of $\mathbb{1}$ in $\mathrm{Symp}(M, \Omega)$. Let $p : P \to S^2$ be a smooth fibration with fiber M endowed with a fiberwise symplectic structure as follows. For each $x \in S^2$, $p^{-1}(x)$ is endowed with a symplectic form Ω_x such that Ω_x depends smoothly on x and $(p^{-1}(x), \Omega_x)$ is symplectomorphic to (M, Ω). In addition, we always choose an orientation on S^2 as part of the data (so P is also oriented). We call $p : P \to S^2$ a *symplectic fibration* (for more details see [MS]).

Every loop $\{f_t\}$ of Hamiltonian diffeomorphisms of M gives rise to a symplectic fibration. Take two copies of the unit 2-disc D_+^2, D_-^2 such that D_+^2 is endowed with the positive orientation and D_-^2 with the reversed one. Define a new manifold

$$P = M \times D_-^2 \cup_\psi M \times D_+^2$$

where ψ is the gluing map

$$\psi : M \times S^1 \to M \times S^1, (z, t) \mapsto (f_t z, t).$$

Clearly P has the natural structure of a symplectic fibration over S^2 since the f_t are symplectomorphisms (and moreover S^2 gets an orientation in the result of this construction).

---

[1]It is not hard to get rid of this assumption, see the theory of *Hamiltonian* symplectic fibrations in [MS] and [P4].

Homotopic loops lead to isomorphic symplectic fibrations i.e. there exist smooth isomorphisms preserving the fiberwise symplectic structure and the orientation. Furthermore, the construction can be reversed. Given a fibration $P \to S^2$ together with a trivialization over one point, one can reconstruct the homotopy class γ. Note also that the class $\gamma = 0$ corresponds to the trivial fibration $S^2 \times (M, \Omega)$. We leave the proof of these statements to the reader. We shall write $P = P(\gamma)$, where γ is the homotopy class of $\{f_t\}$.

Let us visualize the fibration $P(\gamma)$ for the class of the 1-turn rotation of S^2. Note that in this case both the base and the fiber are S^2. It will be useful to identify S^2 with the complex projective line $\mathbb{C}P^1$ and move to the complex setting.

At this point we will make a digression concerning the symplectic geometry of complex projective spaces. Let E be a $2n$-dimensional real vector space endowed with a complex structure j (here j is a linear transformation $E \to E$ with $j^2 = -\mathbb{1}$), a scalar product g and a symplectic form ω such that

$$g(\xi, \eta) = \omega(\xi, j\eta)$$

for all $\xi, \eta \in E$ (cf. 4.1 above). Of course, using j we can consider E as a complex vector space. The pair (ω, g) as above is called a Hermitian structure on the complex space (E, j). Take the unit sphere

$$S = \{\xi \in E \mid g(\xi, \xi) = 1\}.$$

Consider the circle action on S defined by

$$\xi \to e^{2\pi j t}\xi, \ t \in \mathbb{R}/\mathbb{Z}.$$

The orbits of this action are precisely the sets of the form $S \cap l$, where l is a complex line in E. Thus the space of orbits S/S^1 can be canonically identified with the complex projective space $\mathbb{P}(E)$. The action preserves the restriction of ω to TS. Denote by Ω the projection of $\frac{1}{\pi}\omega$ to $\mathbb{P}(E)$. This form is closed. Further, an elementary linear algebra argument shows that Ω is non-degenerate, and thus we get a symplectic form on $\mathbb{P}(E)$.

The form Ω is called the standard (or the Fubini-Study) symplectic form on $\mathbb{P}(E)$ associated to the Hermitian structure (ω, g) on a complex space (E, j). The construction above is a particular case of the Marsden-Weinstein reduction [MS] which plays a fundamental role in the theory of group actions on symplectic manifolds. The factor $\frac{1}{\pi}$ above is chosen for the following reason.

Exercise 9.2.A. Show that the integral of Ω over a projective line in $\mathbb{P}(E)$ equals 1.

It follows from Moser's theorem [MS] that different Hermitian structures on (E, j) give rise to diffeomorphic symplectic forms on $\mathbb{P}(E)$.

Exercise 9.2.B. Consider the space $\mathbb{C}^n$ endowed with the standard Hermitian structure (see 4.1). Show that the standard symplectic form on $\mathbb{P}(\mathbb{C}^n) = \mathbb{C}P^{n-1}$ is invariant under the group $\mathbb{P}U(n) = U(n)/S^1$, and that this group acts on $\mathbb{C}P^{n-1}$ by Hamiltonian diffeomorphisms. Show that

$$\mathbb{P}U(2) = SU(2)/\{\mathbb{1}; -\mathbb{1}\}.$$

Exercise 9.2.C. Consider a loop of projective unitary transformations of $\mathbb{C}P^1$ which in homogeneous coordinates $(z_1 : z_2)$ on $\mathbb{C}P^1$ is given by

$$(z_1 : z_2) \to (e^{-2\pi i t} z_1 : z_2), \ t \in [0; 1].$$

Show that this loop represents the non-trivial element of $\pi_1(\mathrm{Ham}(\mathbb{C}P^1))$. *Hint:* Use the canonical isomorphism $SU(2)/\{\mathbb{1}; -\mathbb{1}\} \to SO(3)$ (see [DFN]).

There exists a natural "parametric" version of the construction above which allows us to produce symplectic fibrations whose fiber is $\mathbb{C}P^{n-1}$ endowed with the standard symplectic form. Let $E \to S^2$ be a complex vector bundle of rank n, and let $\mathbb{P}(E)$ be its projectivization. Every Hermitian structure on E gives rise to the fiber-wise symplectic form Ω_x on $\mathbb{P}(E)$. Here Ω_x equals the standard symplectic form on the fiber $\mathbb{P}(E_x)$. Different choices of the Hermitian structure lead to isomorphic symplectic fibrations.

We return now to the symplectic fibration $P(\gamma)$, where γ is the non-trivial element of $\pi_1(\mathrm{Ham}(S^2))$. Let $T \to \mathbb{C}P^1$ be the tautological line bundle i.e. the one whose fiber over a complex line in $\mathbb{C}^2$ is the line itself. Let $C = \mathbb{C} \times \mathbb{C}P^1$ be the trivial bundle.

Exercise 9.2.D. Prove that the symplectic fibration $P(\gamma)$ is isomorphic to $\mathbb{P}(T \oplus C)$. *Hint:* Represent the base $\mathbb{C}P^1$ as the union of two discs

$$D_- = \{(x_0 : x_1) \in \mathbb{C}P^1 \ \big| \ |x_0/x_1| \le 1\},$$

and

$$D_+ = \{(x_0 : x_1) \in \mathbb{C}P^1 \ \big| \ |x_1/x_0| \le 1\}.$$

Introduce a coordinate t on the circle $S^1 = \partial D_+ = \partial D_-$ by $x_1/x_0 = e^{2\pi i t}$. The bundle $T \oplus C$ can be trivialized over D_- and D_+ in such a way that the transition function $S^1 \times \mathbb{C}^2 \to S^1 \times \mathbb{C}^2$ has the form

$$(t, z_1, z_2) \to (t, e^{-2\pi i t} z_1, z_2)$$

(see e.g. [GH] for the detailed description of the tautological line bundle). The result now follows from 9.2.C above.

Exercise 9.2.E. Show that $\mathbb{P}(T \oplus C)$ is biholomorphically equivalent to the complex blow up of $\mathbb{C}P^2$ at one point. The fibration is obtained by the proper transform of the pencil of lines passing through the blown up point.

9.3 Symplectic connections

Let $p : P \to S^2$ be a symplectic fibration with fiber (M, Ω). A connection σ on P (that is a field of 2-dimensional subspaces transversal to the fibers, see Figure 9), is called symplectic if the parallel transport preserves the fiberwise symplectic structure. One can show that every symplectic fibration admits a symplectic connection, see [GLS], [MS].

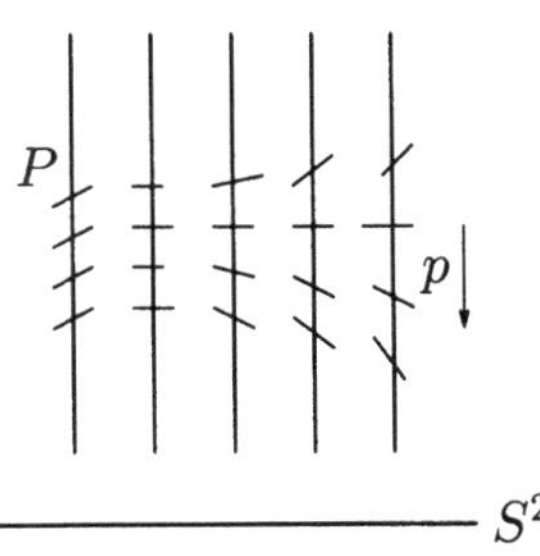

Figure 9

Example. Let $E \to S^2$ be a complex vector bundle endowed with a Hermitian metric. Every Hermitian connection on E induces a symplectic connection on the projectivized bundle $\mathbb{P}(E) \to S^2$.

Recall the definition of curvature of a connection. For $x \in S^2$ and $\xi, \eta \in T_x S^2$, extend ξ and η locally in a neighbourhood of x and consider horizontal lifts $\tilde{\xi}, \tilde{\eta}$ of the obtained vector fields. By definition, the curvature ρ^σ at the point x is a 2-form which takes values in the Lie algebra of vector fields on the fiber $p^{-1}(x)$. So $\rho^\sigma(\xi, \eta) = \left([\tilde{\xi}, \tilde{\eta}] \right)^{\text{vert}}$, where "vert" denotes the projection of $[\tilde{\xi}, \tilde{\eta}]$ onto the fiber $p^{-1}(x)$. Thus if σ is a symplectic connection, then $\rho^\sigma(\xi, \eta)$ lies in the Lie algebra of $\mathrm{Symp}(p^{-1}(x))$, that is $\rho^\sigma(\xi, \eta)$ is a Hamiltonian vector field since $H^1(M, \mathbb{R}) = 0$. Identifying the Hamiltonian vector fields with the normalized Hamiltonian functions, we shall consider $\rho^\sigma(\xi, \eta)$ as a normalized function

$$p^{-1}(x) \to \mathbb{R}.$$

Fix an area form τ on S^2 such that $\int_{S^2} \tau = 1$ (here we need the orientation of S^2). Since every 2-form on S^2 is a multiple of τ we can write

$$\rho^\sigma = L^\sigma \tau,$$

where L^σ is a function on P.

The theory of symplectic connections has been recently developed by Guillemin, Lerman and Sternberg (see [GLS], [MS]). As fundamental objects of this theory arise the coupling form of a symplectic connection and the coupling class of a symplectic fibration. For a point $(x, z) \in P$ we can write

$$T_{(x,z)}P = T_z p^{-1}(x) \oplus T_x S^2,$$

which is the decomposition with respect to the symplectic connection σ. Define the *coupling form δ^σ* of the connection σ as the 2-form on P given by

$$\delta^\sigma(v \oplus \xi, w \oplus \eta) = \Omega_x(v, w) - \rho^\sigma(\xi, \eta)(z).$$

Here $z \in p^{-1}(x)$, $v, w \in T_z p^{-1}(x)$ and $\xi, \eta \in T_x S^2$, while $\rho^\sigma(\xi, \eta)$ is considered as a function on the fiber $p^{-1}(x)$.

It turns out that the coupling form is closed. Denote by c its cohomology class in $H^2(P, \mathbb{R})$. Obviously the restriction of c to any fiber $p^{-1}(x)$ coincides with the class $[\Omega_x]$ of the symplectic form. Furthermore, one can prove that $c^{n+1} = 0$ where $2n = \dim M$. The next result shows that the class c is uniquely determined by these two properties.

Theorem 9.3.A. *(See [GLS], [MS]) The class c is the unique cohomology class in $H^2(P, \mathbb{R})$ such that $c|_{\text{fiber}} = [\Omega_x]$ and $c^{n+1} = 0$.*

In particular, c is an invariant of the symplectic fibration P which does not depend on the choice of the connection σ. We call c the *coupling class* of P. Detailed proofs of all these results can be found in [GLS], [MS].

The next construction plays an important role in our approach to the length spectrum.

The weak coupling construction. *([GLS], [MS]) For $\varepsilon > 0$ sufficiently small there exists a smooth family of closed 2-forms ω_t on P with $t \in [0, \varepsilon)$ such that*

- $\omega_0 = p^* \tau$
- $[\omega_t] = tc + p^*[\tau]$
- $\omega_t|_{\text{fiber}} = t\Omega_x$
- ω_t *is symplectic for all $t > 0$.*

Set $\varepsilon(P) = \sup \varepsilon$, where the supremum is taken over all such deformations. This quantity measures how strong the weak coupling can be. Note that for a trivial fibration, $P = M \times S^2$, $\varepsilon(P) = +\infty$. Thus in a sense $\varepsilon(P)$ measures the non-triviality of the fibration.

There is another way to measure the non-triviality, which works for fibrations with other structure groups as well. It was suggested by Gromov [G2] for unitary vector bundles and we will here give a symplectic version. The idea is to measure the minimal possible norm of the curvature of a symplectic connection on P. We will introduce the following concept.

Definition.

$$\chi_+(P) = \sup_\sigma \frac{1}{\max_P L^\sigma}$$

is the (positive part of the) *symplectic K-area* of P. Here the supremum is taken over all symplectic connections on P, and L^σ is defined by $\rho^\sigma = L^\sigma \tau$.

Exercise. Prove that both quantities introduced above, $\varepsilon(P)$ and $\chi_+(P)$, do not depend on the particular choice of the area form τ on S^2 with $\int_{S^2} \tau = 1$. *Hint:* First apply Moser's theorem [MS] which states that given any two such forms, say τ_1 and τ_2, there exists a diffeomorphism a of S^2 isotopic to the identity such that $a^* \tau_2 = \tau_1$. Then lift a to a fiber-wise symplectic diffeomorphism A of P, which means that $p(A(z)) = a(p(z))$ for all $z \in P$ and the restriction A_x of A to any fiber $p^{-1}(x)$ satisfies $(A_x)^* \Omega_{a(x)} = \Omega_x$. Such a lift can be constructed with the help of any symplectic connection on P. Note that A takes any weak coupling deformation associated to τ_2 to a weak coupling deformation associated to τ_1. This proves that $\varepsilon(P)$ does not depend on the choice of the area form. Further, A acts on the space of symplectic connections on P by $\sigma \to A_* \sigma$. Show that

$$\rho^{A_* \sigma}(a_* \xi, a_* \eta)(Az) = \rho^\sigma(\xi, \eta)(z)$$

for every point $z \in P$ and every pair of vectors $\xi, \eta \in T_{p(z)} S^2$. The fact that $\chi_+(P)$ does not depend on the choice of the area form is a simple consequence of this formula.

Theorem 9.3.B. *([P4]) Let $P = P(\gamma)$. Then $\varepsilon(P) = \chi_+(P) = \frac{1}{\nu_+(\gamma)}$.*

We will prove a weaker statement, namely that $\varepsilon(P) \geq \chi_+(P) \geq \frac{1}{\nu_+(\gamma)}$, but this will be enough to prove Theorem 9.1.A.

Proof of $\varepsilon(P) \geq \chi_+(P)$. Let σ be a symplectic connection. Consider

$$\omega_t = p^* \tau + t\delta^\sigma,$$

where δ^σ is the coupling form. At the point $(x, z) \in P$, we have

$$\omega_{t,(x,z)} = t\Omega_x \oplus -tL^\sigma(x,z)\tau + \tau = t\Omega_x \oplus (1 - tL^\sigma(x,z))\tau.$$

Clearly, ω_t satisfies the first three properties of the weak coupling construction. The form ω_t is symplectic when $1 - tL^\sigma(x,z) > 0$ or equivalently $L^\sigma(x,z) < \frac{1}{t}$ for all x, z. This condition means that $\max_P L^\sigma(x,z) < \frac{1}{t}$, or

$$\frac{1}{\max_P L^\sigma(x,z)} > t.$$

Choose now an arbitrary $\kappa > 0$ and a symplectic connection σ such that

$$\frac{1}{\max\limits_{P} L^\sigma(x,z)} > \chi_+(P) - \kappa.$$

Thus there exists a coupling deformation ω_t for $t \in [0, \chi_+(P) - \kappa)$. That is for all $\kappa > 0$, $\varepsilon(P) \geq \chi_+(P) - \kappa$ and we conclude that

$$\varepsilon(P) \geq \chi_+(P). \qquad \qquad \square$$

Note, that we proved the existence of the weak coupling deformation.

Proof of $\chi_+(P) \geq \frac{1}{\nu_+(\gamma)}$. The proof is based on the following exercise.

Exercise. Let $p : P \to S^2$ be a symplectic fibration. Let ω be a closed 2-form on P such that $\omega|_{\text{fiber}} = \Omega_x$. Set

$$\sigma_{(x,z)} = \{\xi \in T_{(x,z)}P \mid i_\xi \omega = 0 \text{ on } T_z p^{-1}(x)\}.$$

Show that σ defines a symplectic connection on P.

Let $\{f_t\}, t \in [0,1]$, be an arbitrary loop of Hamiltonian diffeomorphisms generated by a normalized Hamiltonian $F \in \mathcal{H}$. Fix polar coordinates $u \in (0,1]$ (the radius) and $t \in S^1 = \mathbb{R}/\mathbb{Z}$ (the normalized angle) on D^2. Take a monotone cut off function $\phi(u)$ such that $\phi(u) = 0$ near $u = 0$ and $\phi(u) = 1$ near $u = 1$. Set

$$P = M \times D^2_- \cup_\psi M \times D^2_+,$$

where $\psi(z,t) = (f_t z, t)$. Define a closed 2-form ω on P by

$$\omega = \begin{cases} \Omega & \text{on } M \times D^2_+ \\ \Omega + d(\phi(u)H_t(z)) \wedge dt & \text{on } M \times D^2_- \end{cases}$$

where $H_t(z) = F(f_t z, t)$.

Exercise. Prove that ω is well defined i.e. show that $\psi^*\Omega = \Omega + dH_t \wedge dt$ (this can be done by a direct computation).

Let us compute the curvature of ρ^σ of the symplectic connection associated to ω. Notice that ρ^σ vanishes on D^2_+ (since Ω induces a flat connection on D^2_+), so it remains to compute ρ^σ on D^2_-. Note that $\rho^\sigma = 0$ near 0 on D^2_-, which is good since we avoid the singularity at zero. In order to compute the curvature, we must find the horizontal lifts of $\frac{\partial}{\partial u}$ and $\frac{\partial}{\partial t}$ at $(x,z) \in M \times D^2_-$.

Let the horizontal lift of $\frac{\partial}{\partial u}$ be of the form

$$\widetilde{\frac{\partial}{\partial u}} = \frac{\partial}{\partial u} + v \text{ for some } v \in T_z p^{-1}(x).$$

Since by definition of the connection, we have $\omega(\widetilde{\frac{\partial}{\partial u}}, w) = 0$ for all $w \in T_z M$, we have

$$0 = \omega\left(\widetilde{\frac{\partial}{\partial u}}, w\right) = \omega(v, w) = \Omega(v, w)$$

for all $w \in T_z p^{-1}(x)$. It follows from the non-degeneracy of Ω that $v = 0$ and hence that

$$\widetilde{\frac{\partial}{\partial u}} = \frac{\partial}{\partial u}.$$

Setting

$$\widetilde{\frac{\partial}{\partial t}} = \frac{\partial}{\partial t} + v \text{ for some other } v \in T_z p^{-1}(x),$$

we get as before, that for all $w \in T_z p^{-1}(x)$

$$0 = \omega\left(\widetilde{\frac{\partial}{\partial t}}, w\right) = \omega(\frac{\partial}{\partial t} + v, w) = \Omega(v, w) - d(\phi(u)H_t)(w).$$

So $i_v \Omega = d(\phi(u)H_t)$ and therefore $v = -\text{ sgrad } \phi(u)H_t = -\phi(u) \text{ sgrad } H_t$ and we can deduce that

$$\widetilde{\frac{\partial}{\partial t}} = \frac{\partial}{\partial t} - \phi(u) \text{ sgrad } H_t.$$

Computing the curvature we get

$$\rho^\sigma\left(\frac{\partial}{\partial t}, \frac{\partial}{\partial u}\right) = \left[\frac{\partial}{\partial t} - \phi(u) \text{ sgrad } H_t, \frac{\partial}{\partial u}\right]^{\text{vert}} = \phi'(u) \text{ sgrad } H_t$$

(actually the commutator is already a vertical vector field). Passing from vector fields to Hamiltonian functions in the definition of curvature we see that

$$\rho^\sigma\left(\frac{\partial}{\partial t}, \frac{\partial}{\partial u}\right) = \phi'(u)H_t(z).$$

Fix $\kappa > 0$ and choose the area form on D^2_- to be $(1-\kappa)dt \wedge du$ (remember that D^2_- has the reversed orientation).

Extend it to an area form on D^2_+ such that Area $(D^2_+) = \kappa$. Denote the resulting area form by τ. We see that $\rho^\sigma = L^\sigma(u, t, z)\tau$, where

$$L^\sigma(u, t, z) = \begin{cases} 0 & \text{on } M \times D^2_+ \\ \frac{\phi'(u)H_t(z)}{1-\kappa} & \text{on } M \times D^2_-. \end{cases}$$

Finally, choose ϕ such that $\phi'(u) \leq 1 + \kappa$ and $\{f_t\}$ such that $\max_z F_t = \max_z H_t \leq \nu_+(\gamma) + \kappa$. We have

$$\max_P L^\sigma \leq \frac{1+\kappa}{1-\kappa}(\nu_+(\gamma) + \kappa)$$

so

$$\chi_+(P) = \sup_\sigma \frac{1}{\max_P L^\sigma}$$

$$\geq \frac{1-\kappa}{(1+\kappa)(\nu_+(\gamma)+\kappa)}.$$

Since κ was arbitrary we have proved that

$$\chi_+(P) \geq \frac{1}{\nu_+(\gamma)}. \qquad \square$$

9.4 An application to length spectrum

The final step in the proof of Theorem 9.1.A is the following estimate for $\varepsilon(P)$ which we will discuss in the next chapter.

Theorem 9.4.A. *Let* $P = P(\gamma)$, *where* γ *is the 1-turn rotation of* S^2. *Then*

$$\varepsilon(P) \leq 2.$$

Proof of 9.1.A. We have seen that $\nu_+(\gamma) \leq \frac{1}{2}$. On the other hand, 9.3.B and 9.4.A imply

$$2 \geq \varepsilon(P) \geq \chi_+(P) \geq \frac{1}{\nu_+(\gamma)},$$

so $\nu_+(\gamma) = \frac{1}{2}$ and $\varepsilon(P) = \chi_+(P) = 2$. $\qquad \square$

Exercise. Let C and T be the trivial and the tautological holomorphic line bundles over $\mathbb{C}P^1$ respectively (see 9.2). Let ∇_C be the natural flat connection on C. There exists a unique connection, say ∇_T, on T which preserves the canonical Hermitian structure on T coming from $\mathbb{C}^2$, and such that its $(0,1)$-part coincides with the $\bar\partial$-operator (see [GH]). Consider symplectic fibration $P = \mathbb{P}(T \oplus C)$ and denote by σ the connection on P which comes from $\nabla_T \oplus \nabla_C$. Obviously, σ is symplectic (actually, parallel transport preserves both the symplectic and the complex structure on the fibers). Let τ be the Fubini-Study area form on $\mathbb{C}P^1$ defined in 9.2 above. Prove that

$$\frac{1}{\max_P L^\sigma} = 2,$$

where $\rho^\sigma = L^\sigma \tau$. In particular, σ is a connection with minimal possible curvature, where the "size" of the curvature is measured with the help of the Fubini-Study area form τ.

Chapter 10

Deformations of Symplectic Forms and Pseudo-holomorphic Curves

In the present chapter we prove the upper bound 9.4.A for the coupling parameter and therefore complete the calculation of the length spectrum of $\mathrm{Ham}(S^2)$. This turns out to be a particular case of a more general problem concerning deformations of symplectic forms (see [P7]). We discuss an approach to the deformation problem which is based on Gromov's theory of pseudo-holomorphic curves.

10.1 The deformation problem

Let (P, ω) be a closed symplectic manifold and l a ray in $H^2(P, \mathbb{R})$ with origin at $[\omega]$, see Figure 10.

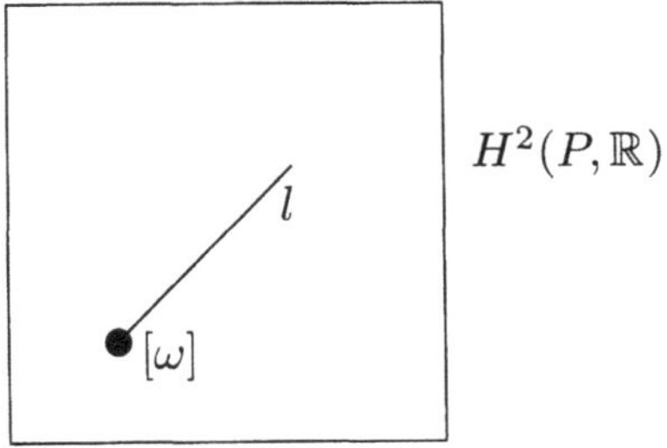

Figure 10

Problem. *How far can one deform ω through symplectic structures on P in such a way that the cohomology class moves along l?*

Note that adding a small closed 2-form to ω leaves it symplectic, but how far can one go? In the case of the coupling deformation, the form $[p^*\tau]$ itself is degenerate, but for small t, $[p^*\tau] + tc$ is already symplectic. In this situation we can go as far along the ray in the direction of the coupling class c as $\varepsilon(P)$.

Here is an example of an obstruction to going infinitely far. Assume that $\dim P = 4$ and that $\Sigma \subset P$ is an embedded 2-sphere such that $\omega|_{T\Sigma}$ is an area form (in other words Σ is a symplectic submanifold of P). We write (A, B) for the intersection index of two homology classes A and B.

Definition. Let $\Sigma \subset P^4$ be a symplectic embedded sphere. If $([\Sigma], [\Sigma]) = -1$ we call Σ an *exceptional sphere*.

Theorem 10.1.A. *([McD1]) Let $\Sigma \subset (P^4, \omega)$ be an exceptional sphere. Let $\omega_t, t \in [0,1]$ be a deformation of ω through symplectic forms. Then $([\omega_1], [\Sigma]) > 0$.*

In other words, the hyperplane $(x, [\Sigma]) = 0$ in $H^2(P, \mathbb{R})$ is a wall which cannot be crossed by a symplectic deformation.

The proof is based on the theory of pseudo-holomorphic curves. Below we give a sketch of the proof of this statement as well as an application to the proof of 9.4.A.

10.2 The $\bar\partial$-equation revisited

An almost complex structure j on a manifold P is a field of endomorphisms $TP \to TP$ such that $j^2 = -\mathbb{1}$. An important class of examples comes from complex algebraic geometry. Every complex manifold (that is a manifold equipped with an atlas whose transition functions are holomorphic) has the canonical almost complex structure $\xi \to \sqrt{-1}\xi$. Almost complex structures arising in this way are called *integrable*. A deep fact is that integrability is equivalent to the vanishing of a certain tensor associated to the almost complex structure ([NN]). As a consequence of this result one gets that every almost complex structure on a (real) surface is integrable. Moreover, on the 2-sphere all almost complex structures are diffeomorphic (a classical fact called the uniformization theorem [AS]).

Definition. Let (P, ω) be a symplectic manifold. An almost complex structure j is *compatible with* ω if $g(\xi, \eta) = \omega(\xi, j\eta)$ defines a Riemannian metric on P.

Exercise 10.2.A. Let (P, ω) be a real surface and let j be an almost complex structure on P. Then either j or $-j$ is compatible with ω.

Let (P, ω) be a symplectic manifold, and let j be a compatible integrable almost complex structure on P. The triple (P, ω, j) is called a *Kähler* structure. For instance, $\mathbb{C}P^n$ endowed with the standard ω and j is a Kähler manifold (cf. 9.2 above). Thus every complex submanifold of $\mathbb{C}P^n$ is Kähler with respect to the induced structure. Vice versa, if (P, ω, j) is a closed Kähler manifold such that the cohomology class $[\omega]$ is integral, then (P, j) admits a holomorphic embedding into $\mathbb{C}P^n$ for some n (this is the famous Kodaira embedding theorem, see [GH]).

It was Gromov's great insight [G1] that one can generalize some important methods in algebraic geometry to *quasi-Kähler manifolds* (P, ω, j). Here j is a compatible almost complex structure but in general not integrable anymore. Remarkably enough, the theory of holomorphic curves extends without essential changes to the non-integrable case. The importance of such a generalization is due to the following reason. Every symplectic manifold admits a compatible almost complex structure. However there are symplectic manifolds which cannot have a Kähler structure (see [MS]).

Exercise 10.2.B (see [MS]). Let E be an even dimensional linear space endowed with a non-degenerate skew-symmetric bilinear form ω. Show that the space of complex structures $j : E \to E$, $j^2 = -\mathbb{1}$ is contractible.

Therefore a compatible almost complex structure on a symplectic manifold is a section of a fiber bundle whose fibers are contractible. Hence such structures do exist and moreover form a contractible space. By the same reason various extension problems connected to almost complex structures admit a positive solution.

Exercise 10.2.C. Let (P, ω) be a 4-dimensional symplectic manifold, and let $\Sigma \subset P$ be a symplectic submanifold. Assume that Σ is equipped with an almost complex structure j which is compatible with $\omega|_{T\Sigma}$. Show that j extends to a compatible almost complex structure on P.

Let us turn now to the theory of pseudo-holomorphic curves on quasi-Kähler manifolds.

Definition. A map $\phi : (S^2, i) \to (P, j)$ is a *pseudo-holomorphic (or j-holomorphic) curve* if $\phi_* \circ i = j \circ \phi_*$.

Exercise 10.2.D. Show that on a complex manifold the definition above is equivalent to the usual Cauchy-Riemann equation.

We set

$$\bar{\partial}\phi = \frac{1}{2}(\phi_* + j \circ \phi_* \circ i).$$

Exercise 10.2.E(cf. 4.1.A). Given (P, ω, j, g) and a j-holomorphic curve $\phi : (S^2, i) \to (P, j)$, show that $\mathrm{Area}_g(\phi(S^2)) = \int_{S^2} \phi^*\omega$. In particular, if ϕ is non-constant then $\int_{S^2} \phi^*\omega > 0$.

Moreover, if ϕ is an embedding, then $\phi(S^2)$ is a symplectic submanifold of P. Indeed, the restriction of the symplectic form coincides with the Riemannian area form.

Let (P, ω) be a symplectic manifold. Choose an almost complex structure j compatible with ω. Then the tangent bundle TP gets the structure of a complex vector bundle. Since the space of compatible j's is connected the corresponding characteristic classes do not depend on the choice of j. We write c_1 for the first Chern class of TP with respect to any compatible almost complex structure.

Exercise 10.2.F (Adjunction formula). Let (P^4, ω) be a symplectic manifold with a compatible almost complex structure j. Let $\Sigma \subset P$ be an embedded j-holomorphic sphere. Show that

$$1 + \frac{1}{2}(([\Sigma], [\Sigma]) - c_1(\Sigma)) = 0.$$

Hint: Use that $([\Sigma], [\Sigma])$ is the self-intersection in the complex normal bundle ν_Σ and $T_\Sigma P = T\Sigma \oplus \nu_\Sigma$.

Corollary. *For a symplectically embedded sphere Σ, $c_1 = 1$ if and only if $([\Sigma], [\Sigma]) = -1$.*

Proof. Since Σ is symplectically embedded, there exists an ω-compatible almost complex structure j such that Σ is j-holomorphic (see 10.2.C). The statement now follows from the adjunction formula 10.2.F. $\square$

10.3   An application to coupling

In this section we deduce 9.4.A from 10.1.A. Recall that we study the coupling deformation for the fibration $\mathbb{P}(T \oplus C) \to \mathbb{C}P^1$, where T and C are the tautological and trivial bundle respectively.

Exercise. Let E be a complex vector space and let $l \in \mathbb{P}(E)$ be a line in E. Show that $T_l\mathbb{P}(E)$ is canonically isomorphic to $\mathrm{Hom}(l, E/l) = l^* \otimes E/l$.

First of all, we wish to compute the (co)homology ring of $P = \mathbb{P}(T \oplus C)$. Denote by $[F]$ the homology class of the fiber and by Σ the section corresponding to the rank 1 subbundle $0 \oplus C$. Clearly, $([F], [F]) = 0$ and $([F], [\Sigma]) = 1$. In order to compute $([\Sigma], [\Sigma])$, notice that the normal bundle ν_Σ is just the restriction on Σ of the tangent bundle to the fibers. The above exercise implies that

$$\nu_\Sigma = \mathrm{Hom}(C, T \oplus C/C) = \mathrm{Hom}(C, T) = C^* \otimes T = T.$$

Thus $c_1(\nu_\Sigma) = -1$. We conclude that $([\Sigma], [\Sigma]) = -1$ since$([\Sigma], [\Sigma])$ is just the self-intersection in the normal bundle.

Let ω_t be a coupling deformation. It is an easy consequence of Moser's theorem [MS] that for small enough t, ω_t is symplectomorphic to a Kähler form with respect to the standard complex structure on $\mathbb{P}(T \oplus C)$. Without loss of generality, we may assume that ω_t is a Kähler form for small t. Since Σ is a holomorphic section of P, we get that Σ is symplectic and embedded. Taking into account that $([\Sigma], [\Sigma]) = -1$ we see that Σ is an exceptional sphere. Thus Theorem 10.1.A implies that $([\omega_t], [\Sigma]) > 0$ for all t.

Recall, that $[\omega_t] = p^*[\tau] + tc$, where $[\tau]$ is a generator of $H^2(S^2, \mathbb{Z})$ coming from the orientation and c is the coupling class. The classes $p^*[\tau]$ and c are uniquely defined by the following relations:

$$(p^*[\tau], [F]) = 0 \qquad (p^*[\tau], [\Sigma]) = 1$$
$$(c, [F]) = 1 \qquad c^2 = 0.$$

We shall use Poincaré duality to identify homology and cohomology.

Exercise. Prove that $p^*[\tau] = [F]$ and $c = [\Sigma] + \frac{1}{2}[F]$.

Thus, in view of Theorem 10.1.A

$$([\omega_t], [\Sigma]) = ([F] + t[\Sigma] + \frac{t}{2}[F], [\Sigma]) = 1 - t + \frac{t}{2} = 1 - \frac{t}{2} > 0,$$

which implies that $t < 2$. Since this is true for any coupling deformation, we have proved that $\varepsilon(P) \leq 2$. This completes the proof. $\qquad\square$

10.4 Some properties of pseudo-holomorphic curves in symplectic manifolds

Here we outline Gromov's theory of pseudo-holomorphic curves (see [G1], [AL]). Let (P^{2n}, ω) be a symplectic manifold and let $A \in H_2(P, \mathbb{Z})$ be a primitive class. This means that A cannot be represented as kB where $k > 1$ is an integer and $B \in H_2(P, \mathbb{Z})$. In particular, $A \neq 0$. Let $\mathcal{J}$ be the space of all ω-compatible almost complex structures on P and let $\mathcal{N}$ be the space of all smooth maps $f : S^2 \to P$ such that $[f] = A$. Define $\mathcal{X} \subset \mathcal{N} \times \mathcal{J}$ by

$$\mathcal{X} = \{(f, j) | f \in \mathcal{N}, j \in \mathcal{J} \text{ and } \bar{\partial}_j f = 0\}.$$

Then $\mathcal{X}$ is a smooth submanifold of $\mathcal{N} \times \mathcal{J}$ and the projection $\pi : \mathcal{X} \to \mathcal{J}$ is a Fredholm operator i.e. Ker π_* and Coker π_* are finite dimensional. Its index satisfies

$$\text{Index } \pi_* := \dim (\text{Ker } \pi_*) - \dim (\text{Coker } \pi_*) = 2(c_1(A) + n).$$

The Fredholm property of π allows us to apply an infinite dimensional version of Sard's theorem [Sm].

Let j_0 and j_1 be regular values of π (i.e π_* is surjective for all $x \in \pi^{-1}(j_k), k = 0, 1$). So $\pi^{-1}(j_0)$ and $\pi^{-1}(j_1)$ are smooth submanifolds. Then for a generic path γ joining j_0 to j_1, $\pi^{-1}(\gamma)$ is a smooth submanifold of dimension Index $(\pi) + 1$ and $\partial \pi^{-1}(\gamma) = \pi^{-1}(j_0) \cup \pi^{-1}(j_1)$ (see Figure 11). The crucial role is played by the compactness properties of $\pi^{-1}(\gamma)$.

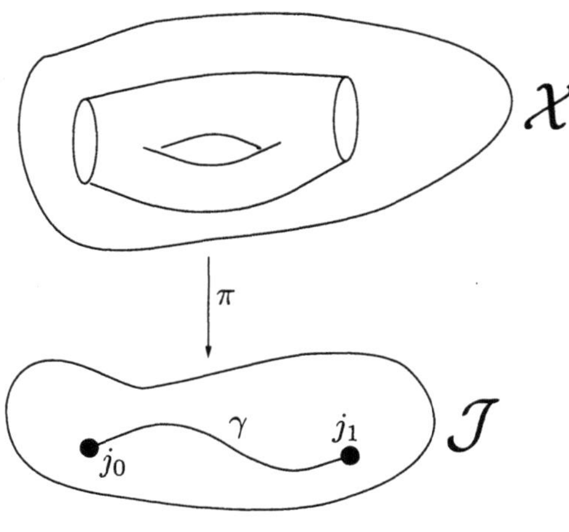

Figure 11

First of all, notice that $\pi^{-1}(\gamma)$ cannot be compact itself since it admits the action of the non-compact group $PSL(2, \mathbb{C})$. Here $PSL(2, \mathbb{C})$ is the group of conformal transformations of (S^2, i) and it acts as follows. If $h : S^2 \to S^2$ is a conformal transformation and $(f, j_0) \in \pi^{-1}(j_0)$ then we also have $(f \circ h, j_0) \in \pi^{-1}(j_0)$.

Exercise. Show that this action is free (use that A is primitive).

Gromov's compactness theorem. *Either the moduli space $\pi^{-1}(\gamma)/PSL(2, \mathbb{C})$ is compact or there exists a family $(f_k, j_k) \in \pi^{-1}(\gamma)$ such that $j_k \to j_\infty$ and f_k "converges" to a j_∞-holomorphic cusp curve in the class A.*

We do not specify the convergence here. What is important to us is that if $\pi^{-1}(\gamma)/PSL(2, \mathbb{C})$ is not compact, then there exists a j-holomorphic cusp curve in the class A for some $j \in \mathcal{J}$. A cusp curve is defined as follows. Let $A = A_1 + \cdots + A_d$ for $d > 1$ be a decomposition of A such that $A_k \neq 0$ for all k and let $\phi_k : S^2 \to P$ be j-holomophic curves in the classes $A_k, k = 1, \ldots, d$. We say that these data define a cusp curve in the class A. The union $\bigcup \phi_k(A_k)$ is called its image and is usually assumed to be connected.

Imagine now the following situation.

- The set $\mathcal{J}'$ of those $j \in \mathcal{J}$, which have cusp curves in the class A, has codimension at least 2 in $\mathcal{J}$.
- $j_0 \in \mathcal{J} \setminus \mathcal{J}'$ is a regular value of π and $\pi^{-1}(j_0)/PSL(2,\mathbb{C})$ (which is thus a compact manifold without boundary) is not cobordant to zero (i.e. it does not bound a compact manifold).

For instance, a point is not cobordant to zero, since it does not bound a compact manifold.

Then for every regular value $j_1 \in \mathcal{J} \setminus \mathcal{J}'$ the set $\pi^{-1}(j_1)/PSL(2,\mathbb{C})$ is not empty. Indeed, if it were empty, and we joined j_0 to j_1 by a generic path $\gamma \in \mathcal{J} \setminus \mathcal{J}'$, we would have that

$$\partial\big(\pi^{-1}(\gamma)/PSL(2,\mathbb{C})\big) = \pi^{-1}(j_0)/PSL(2,\mathbb{C}).$$

This contradicts the fact that $\pi^{-1}(j_0)/PSL(2,\mathbb{C})$ is not cobordant to zero, see Figure 12 which illustrates the contradiction.

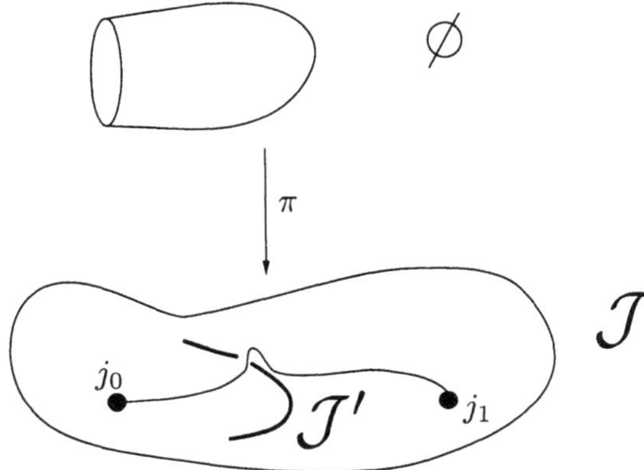

Figure 12

This reflects the persistence principle, which we already encountered in the case of pseudo-holomorphic discs in 4.2. Either solutions of the $\bar{\partial}$-equation persist or bubbling off takes place.

10.5   Persistence of exceptional spheres

Here we sketch the proof of Theorem 10.1.A. Let (P^4, ω) be a symplectic manifold and let $\Sigma \subset P$ be an exceptional sphere with $A = [\Sigma]$. Let $\omega_t, t \in [0,1]$ be a deformation of the symplectic form $\omega = \omega_0$.

Step 1. Choose an ω-compatible almost complex structure j_0 such that Σ is j_0-holomorphic and extend j_0 to a one-parameter family j_t such that j_t is ω_t-compatible (cf. 10.2.C above).

Step 2. The theory described in the previous section holds without changes for the set of j's which are compatible with symplectic structures in a given deformation class. So

$$\text{Index } \pi = 2(c_1(A) + n) = 2(1 + 2) = 6.$$

However, $\dim {}_{\mathbb{R}} PSL(2,\mathbb{C}) = 6$ so $\dim \left(\pi^{-1}(j_0)/PSL(2,\mathbb{C})\right) = 0$ provided j_0 is regular.

Step 3. In dimension 4 we have the following phenomenon. Two different germs of j-holomorphic curves always intersect with positive index at a common point. This fact is well known for integrable j's. In the non-integrable case this is a simple linear algebra statement provided the intersection is transversal. However the proof requires a delicate local analysis for non-transversal intersections. As a corollary, there is a unique j_0-holomorphic curve in A. Since if there were another one, say Σ', then $([\Sigma], [\Sigma']) = ([\Sigma], [\Sigma]) = -1$ contradicting that $([\Sigma], [\Sigma']) \geq 0$. We conclude that $\pi^{-1}(j_0)/PSL(2,\mathbb{C})$ consists of exactly one point.

Step 4. Choose j_0 and j_t to be regular. We claim that generically no cusp curves appear. Indeed, set $A = A_1 + \cdots + A_d$, $d > 1$ and

$$c_1(A) = 1 = c_1(A_1) + \cdots + c_1(A_d).$$

Thus at least one of the Chern classes on the right-hand side is non-positive. Assume without loss of generality, that $c_1(A_1) \leq 0$ and that A_1 is represented by a j_s-holomorphic curve for some $s \in [0,1]$. Denote by π_{A_1} the projection in the class A_1. We see that $\pi_{A_1}^{-1}(\gamma)/PSL(2,\mathbb{C})$ is non-empty. On the other hand,

$$\dim \pi_{A_1}^{-1}(\gamma)/PSL(2,\mathbb{C}) = 2(c_1(A_1) + 2) - 6 + 1 \leq -1,$$

which is impossible. Thus generically no bubbling off happens (this reflects the fact that $\text{codim } \mathcal{J}' \geq 2$). Hence A is represented by a j_1-holomorphic curve and therefore (see 10.2.E) $([\omega_1], [A]) > 0$. $\square$

Chapter 11

An Application to Ergodic Theory

In the present chapter we discuss an asymptotic geometric invariant associated to the fundamental group of $\mathrm{Ham}(M,\Omega)$ and describe an application to classical ergodic theory (see [P9]).

11.1 Hamiltonian loops as dynamical objects

Let (M,Ω) be a closed symplectic manifold. Given an irrational number α and a smooth loop $h : S^1 \to \mathrm{Ham}(M,\Omega)$, one can define *a skew product map* $T_{h,\alpha} : M \times S^1 \to M \times S^1$ by $T_{h,\alpha}(y,t) = (h(t)y, t + \alpha)$. Our purpose is to relate the geometry and topology of Hamiltonian loops to the dynamics of the associated skew products.[1]

The definition above is a particular case of a much more general notion of a skew product (see [CFS, p. 231]) which has been intensively studied during several decades. There are at least two important reasons for the interest in this notion. Firstly, it serves as a foundation for mathematical models of random dynamics (see [Ki] for a survey). Secondly, it provides non-trivial examples of systems with interesting dynamical properties.

The dynamical property we consider is the strict ergodicity. Recall that a homeomorphism T of a compact topological space X is called *strictly ergodic* if it has precisely one invariant Borel probability measure, say m, which in addition is positive on non-empty open subsets. Strictly ergodic homeomorphisms are ergodic, and have a number of additional remarkable features. We mention one of them which plays a crucial role below. Namely, given such a T and an arbitrary continuous function F on X, the time averages $\frac{1}{N}\Sigma_{i=0}^{N-1}F(T^i x)$ converge uniformly to the space average $\int_X F\,dm$, and in particular converge *for*

---

[1] In this chapter we work with free loops and do not assume that $h(0) = \mathbf{1}$.

all $x \in X$. Note that for general ergodic transformations such a convergence takes place only for *almost all x*. The contrast between "all" and "almost all" becomes especially transparent when one notices that there are purely topological obstructions to the strict ergodicity. For instance, the 2-sphere admits no strictly ergodic homeomorphisms. Indeed the Lefschetz theorem implies that every homeomorphism of S^2 has either a fixed point, or a periodic orbit of period 2 and we see that the invariant measure which is concentrated on such an orbit contradicts the definition of strict ergodicity. In 11.2 below we describe a more sophisticated obstruction to the strict ergodicity which comes from symplectic topology.

We say that a loop $h : S^1 \to \mathrm{Ham}(M, \Omega)$ is *strictly ergodic* if for some α the corresponding skew product map $T_{h,\alpha}$ is strictly ergodic.[2] With this language our central question can be formulated as follows.

Question. Which homotopy classes $S^1 \to \mathrm{Ham}(M, \Omega)$ can be represented by strictly ergodic loops?

Here is an example where one gets a complete answer to this question. Let M_* be the blow up of the complex projective plane $\mathbb{C}P^2$ at one point. Choose a Kähler symplectic structure Ω_* on M_* which integrates to 1 over a general line and to $\frac{1}{3}$ over the exceptional divisor. In view of 9.2.E above one can equivalently think that $M_* = \mathbb{P}(T \oplus C)$, where T and C are the tautological and the trivial holomorphic line bundles over $\mathbb{C}P^1$ respectively. With this language the exceptional divisor is our old friend Σ from 10.3, while the general line is homologous to the sum of Σ with the fiber. The periods of the symplectic form are chosen in such a way that its cohomology class is a multiple of the first Chern class of M (see Definition 11.3.A below). One can easily see that (M_*, Ω_*) admits an effective Hamiltonian action of the unitary group $U(2)$, in other words there exists a monomorphism $i : U(2) \to \mathrm{Ham}(M_*, \Omega_*)$. The fundamental group of $U(2)$ equals $\mathbb{Z}$. It was proved recently by Abreu and McDuff [AM] that the inclusion $\pi_1(U(2)) \to \pi_1(\mathrm{Ham}(M_*, \Omega_*))$ is an isomorphism, and thus $\pi_1(\mathrm{Ham}(M_*, \Omega_*)) = \mathbb{Z}$. As far as I know this is the simplest example of a symplectic manifold with $\pi_1(\mathrm{Ham}) = \mathbb{Z}$.

Theorem 11.1.A. *The trivial class* $0 \in \pi_1(\mathrm{Ham}(M_*, \Omega_*))$ *is the only one which can be represented by a strictly ergodic loop.*

The proof of this theorem consists of two parts. First of all, one has to establish existence of contractible strictly ergodic loops. This can be done by purely ergodic methods in a fairly general situation. We refer the reader to [P9] for an extensive discussion and proofs. Secondly, one has to prove that every class

---

[2]Note that each $T_{h,\alpha}$ preserves the canonical measure on $M \times S^1$ induced by the symplectic form. Thus in our setting the strict ergodicity means that this measure is (up to a factor) the unique invariant measure.

$\gamma \neq 0$ cannot be represented by a strictly ergodic loop. It turns out that the obstruction comes from the geometry of $\mathrm{Ham}(M, \Omega)$. We are going to discuss this in more details.

11.2 The asymptotic length spectrum

Define *the asymptotic norm* of an element $\gamma \in \pi_1(\mathrm{Ham}(M, \Omega))$ by

$$\nu_\infty(\gamma) = \lim_{k \to +\infty} \frac{\nu(k\gamma)}{k},$$

where ν is the norm introduced in 7.3. This notion is similar to asymptotic growth μ defined in 8.2 above. The limit exists since the sequence $\nu(k\gamma)$ is subadditive.

Theorem 11.2.A. *Let $\gamma \in \pi_1(\mathrm{Ham}(M, \Omega))$ be a class represented by a smooth strictly ergodic loop. Then the asymptotic norm $\nu_\infty(\gamma)$ vanishes.*

Proof. The proof is based on an asymptotic curve shortening procedure in the spirit of 8.3 above. Let $h : S^1 \to \mathrm{Ham}(M, \Omega)$ be a smooth loop of Hamiltonian diffeomorphisms which defines a strictly ergodic skew product $T(y, t) = (h(t)y, t+\alpha)$. Let γ be the corresponding element in $\pi_1(\mathrm{Ham}(M, \Omega))$. Denote by $H(x, t)$ the normalized Hamiltonian function generating the loop $h(t)^{-1}$. Set $h_k(t) = h(t + k\alpha)^{-1}$ and set

$$f_N(t) = h_0(t) \circ \cdots \circ h_{N-1}(t).$$

It follows from 1.4.D that the loop f_N is generated by the normalized Hamiltonian function

$$F_N(y, t) = H(y, t) + H(h_0(t)^{-1}y, t + \alpha) + \cdots$$

$$+ H(h_{N-2}(t)^{-1} \circ \cdots \circ h_0(t)^{-1}y, t + (N-1)\alpha).$$

This expression can be rewritten as follows:

$$F_N(y, t) = \sum_{k=0}^{N-1} H \circ T^k(y, t).$$

Since T is strictly ergodic and the function F_N has zero mean we conclude that

$$\frac{1}{N} \int_0^1 \max_{y \in M} F_N(y, t) - \min_{y \in M} F_N(y, t) \, dt \to 0,$$

when $N \to \infty$.

But the expression on the left-hand side is exactly $\frac{1}{N}$ length$\{f_N(t)\}$. Note now that the loop $\{f_N(t)\}$ represents the element $-N\gamma$. Since $\nu(N\gamma) = \nu(-N\gamma)$ we get that $\nu(N\gamma)/N$ tends to zero when $N \to \infty$. This proves that the asymptotic norm of γ vanishes. $\square$

I do not know the *precise* value of $\nu_\infty(\gamma)$ in any example where this quantity is strictly positive (for instance, for the blow up of $\mathbb{C}P^2$ in 11.1 above). The difficulty is the following. In all known examples where Hofer's norm $\nu(\gamma)$ can be computed precisely there exists a closed loop $h(t)$ which minimizes the length in its homotopy class (that is *a minimal closed geodesic*). It turns out however that every non-constant minimal closed geodesic loses minimality after a suitable number of iterations. In other words the loop $h(Nt)$ can be shortened provided N is large enough. The proof of this statement is based on the following generalization of the shortening procedure described above. Let $H(y,t)$ be the normalized Hamiltonian of $h(t)^{-1}$. Assume without loss of generality that $h(0) = \mathbb{1}$, and that $H(y,0)$ does not vanish identically. Denote by Γ the set of all points of M where the function $|H(y,0)|$ attains the maximal value. Since $M \setminus \Gamma$ is a non-empty open subset, and the group of Hamiltonian diffeomorphisms acts transitively on M, one can choose a sequence

$$\phi_0 = \mathbb{1}, \phi_1, \ldots, \phi_{N-1} \in \mathrm{Ham}(M, \Omega)$$

such that

$$\Gamma \cap \phi_1(\Gamma) \cap \cdots \cap \phi_{N-1}(\Gamma) = \emptyset.$$

Consider the loop $f_N(t) = h(t)^{-1} \circ \phi_1 h(t)^{-1} \phi_1^{-1} \circ \cdots \circ \phi_{N-1} h(t)^{-1} \phi_{N-1}^{-1}$. We claim that it is shorter than the loop $h(Nt)$. Indeed, note that its Hamiltonian F_N at time $t = 0$ can be written as follows:

$$F_N(y, 0) = \sum_{i=0}^{N-1} H(\phi_i^{-1} y, 0).$$

Set

$$a(t) = \max_{y \in M} F_N(y, t) - \min_{y \in M} F_N(y, t)$$

and

$$b(t) = N(\max_{y \in M} H(y, t) - \min_{y \in M} H(y, t)).$$

Our choice of the sequence $\{\phi_i\}$ implies that $a(0) < b(0)$. Since $a(t) \leq b(t)$ for all t, we get that $\int_0^1 a(t)dt < \int_0^1 b(t)dt$, and this proves the claim. We conclude that *if a non-zero class $\gamma \in \pi_1(\mathrm{Ham}(M, \Omega))$ is represented by a minimal closed geodesic then $\nu_\infty(\gamma)$ is strictly less than $\nu(\gamma)$.*

It would be interesting to investigate further restrictions on the homotopy classes of smooth strictly ergodic loops in the group of Hamiltonian diffeomorphisms.

11.3 Geometry via algebra

Let us return to Theorem 11.1.A. In this section we very briefly sketch the proof of the fact that non-contractible loops on $\mathrm{Ham}(M_*, \Omega_*)$ cannot be strictly ergodic.

Let (M^{2n}, Ω) be a closed symplectic manifold. Define a mapping

$$I : \pi_1(\mathrm{Ham}(M, \Omega)) \to \mathbb{R}$$

as follows. Take an element $\gamma \in \pi_1(\mathrm{Ham}(M, \Omega))$, and consider the associated symplectic fibration $P(\gamma)$. Denote by u the first Chern class of the vertical tangent bundle to $P(\gamma)$. The fiber of this bundle at a point of $P(\gamma)$ is the (symplectic) vector space tangent to the fiber through this point. As before, we write c for the coupling class. Define the "characteristic number"

$$I(\gamma) = \int_{P(\gamma)} c^n \cup u.$$

It is easy to see that $I : \pi_1(\mathrm{Ham}(M, \Omega)) \to \mathbb{R}$ is a homomorphism ([P6], [LMP2]).

Definition 11.3.A. A symplectic manifold (M, Ω) is called *monotone* if $[\Omega]$ is a positive multiple of $c_1(TM)$.

Theorem 11.3.B. *([P6]) Let (M, Ω) be a closed monotone symplectic manifold. Then there exists a positive constant $C > 0$ such that $\nu(\gamma) \geq C|I(\gamma)|$ for all $\gamma \in \pi_1(\mathrm{Ham}(M, \Omega))$.*

In other words, the homomorphism I calibrates Hofer's norm on the fundamental group. The proof of this theorem is based on the theory described in two previous chapters combined with results from [Se]. Recently Seidel obtained a generalization of this inequality to non-monotone symplectic manifolds.

Since I is a homomorphism, the estimate in 11.3.B extends to the asymptotic Hofer's norm:

$$\nu_\infty(\gamma) \geq C|I(\gamma)|.$$

The great advantage of the homomorphism I is that it can be relatively easily calculated in examples. For instance, one can show that $I(\gamma) \neq 0$ where γ is the generator of $\pi_1(\mathrm{Ham}(M_*, \Omega_*)) = \mathbb{Z}$. Thus 11.3.B implies that the asymptotic norm of every non-trivial element of $\pi_1(\mathrm{Ham}(M_*, \Omega_*))$ is strictly positive. We conclude from 11.2.A that such an element cannot be represented by a strictly ergodic loop.

Chapter 12

Elements of Variational Theory of Geodesics

We have already seen a number of results concerning length minimizing geodesics on the group of Hamiltonian diffeomorphisms. In the present chapter we discuss a different viewpoint on geodesics based on calculus of variations. Given a smooth path of Hamiltonian diffeomorphisms, is it possible to shorten it by a small variation with fixed endpoints? This question is motivated by the classical theory of geodesics on Riemannian manifolds. Interestingly enough, in Hofer's geometry at least under certain non-degeneracy assumptions, one can give a rather precise answer with a transparent dynamical meaning [U].

12.1 What are geodesics?

The intuition coming from Riemannian geometry suggests that the geodesics should be defined as critical points of the length functional. Let us try to formalize this definition.

Let $\{f_t\}, t \in [a; b]$ be a smooth path in $\mathrm{Ham}(M, \Omega)$. A *variation* of $\{f_t\}$ is a smooth family of paths $\{f_{t,\varepsilon}\}$ with $t \in [a; b]$ and $\varepsilon \in (-\varepsilon_0, \varepsilon_0)$ satisfying

$$f_{a,\varepsilon} = f_a, \quad f_{b,\varepsilon} = f_b \quad \text{and} \quad f_{t,0} = f_t$$

for all t and ε. We always assume that $\bigcup_{t,\varepsilon} \mathrm{supp}\, f_{t,\varepsilon}$ is compact. Given such a variation, consider the length of the path $\{f_{t,\varepsilon}\}$ as the function of ε:

$$\ell(\varepsilon) = \int_a^b \|F(\cdot, t, \varepsilon)\| dt$$

$$= \int_a^b \max_x F(x, t, \varepsilon) - \min_x F(x, t, \varepsilon) dt,$$

where $F(x, t, \varepsilon)$ is the Hamiltonian generating the path $\{f_{t,\varepsilon}\}$ for given ε.

Tentative Definition 12.1.A. A path $\{f_t\}$ is a geodesic if

- it has constant speed, that is $\|F(\cdot, t)\|$ does not depend on t;
- for every smooth variation of $\{f_t\}$ the length function $\ell(\varepsilon)$ has a critical point at $\varepsilon = 0$.

Here we immediately face a difficulty – even for smooth variations the function $\ell(\varepsilon)$ need not be smooth! Thus the notion of a critical point should be specified. Our first task is to figure out the structure of the length functions $\ell(\varepsilon)$ associated to variations of the given path.

Proposition 12.1.B. *Every length function $\ell(\varepsilon)$ is convex at 0 up to second order terms. Namely, there exists a convex function $u(\varepsilon)$ and numbers $\delta > 0, C > 0$ such that*

$$|\ell(\varepsilon) - u(\varepsilon)| \leq C\varepsilon^2$$

for all $\varepsilon \in (-\delta, \delta)$.

Note that whatever the definition of the critical point is, it should not depend on the second order terms. The only natural candidate for the critical point of a convex function is its minimum. Thus we arrive at the following notion.

Definition 12.1.C. The point $\varepsilon = 0$ is a critical point for the length function $\ell(\varepsilon)$ if and only if it is a point of minimum for a convex function $u(\varepsilon)$ which satisfies inequality 12.1.B.

Exercise 12.1.D.

- The definition above is correct, that is it does not depend on the particular choice of a convex function u which satisfies 12.1.B.
- If $\ell(\varepsilon)$ is smooth, Definition 12.1.C coincides with the usual one.
- If ℓ attains its local minimum or maximum at 0 then 0 is a critical point of ℓ in the sense of 12.1.C.
- Conclude from 12.1.B that the germ of the function $1 - |\varepsilon|$ at 0 cannot arise as the length function of any smooth variation.

Now we are ready to answer the question posed in the title of this section: **a path $\{f_t\}$, $t \in [a; b]$ is called a geodesic if it satisfies 12.1.A and 12.1.C.**

It follows from 12.2.A below that the restriction of a geodesic defined on $[a; b]$ to any subsegment is again a geodesic. Thus the notion of geodesic extends in the obvious way to paths which are defined on arbitrary time intervals. Let $I \subset \mathbb{R}$ be an interval, and let $f : I \to \mathrm{Ham}(M, \Omega)$ be a smooth path. If the restriction of f to any segment $[a; b] \subset I$ is a geodesic the path f itself is called a geodesic. We will focus on geodesics $\{f_t\}$ which are defined on the unit interval $[0; 1]$. Moreover since Hofer's metric is biinvariant we can always shift a geodesic and assume that $f_0 = \mathbb{1}$.

Let us turn to the proof of 12.1.B. It is convenient to use the linearization idea introduced in Chapter 5. Consider the space $\mathcal{F}_0$ of all normalized Hamiltonians $M \times [0; 1] \to \mathbb{R}$ endowed with the norm

$$|||F|||_0 = \int_0^1 \max_x F(x,t) - \min_x F(x,t)dt.$$

Denote by $\mathcal{H}_0 \subset \mathcal{F}_0$ the subset consisting of all Hamiltonians which generate a loop $\{h_t\}$ of Hamiltonian diffeomorphisms: $h_0 = h_1 = \mathbb{1}$. Let us emphasize that the functions from $\mathcal{F}_0$ and $\mathcal{H}_0$ are **not** assumed to be periodic in time, in the contrast with the setting of Chapter 5. Denote by $\mathcal{V}$ the set of all smooth families $H(x,t,\varepsilon)$ of functions from $\mathcal{H}_0$ with $H(x,t,0) \equiv 0$.

Proposition 12.1.E. *Let $\{f_t\}$ be a smooth path of Hamiltonian diffeomorphisms with $t \in [0; 1]$ and $f_0 = \mathbb{1}$. The set of length functions $\ell(\varepsilon)$ associated with the variations of $\{f_t\}$ consists of all functions of the form*

$$|||F - H(\varepsilon)|||_0,$$

where $H \in \mathcal{V}$.

Proposition 12.1.B is an immediate consequence of this formula. Indeed, set $u(\varepsilon) = |||F - \varepsilon H'(0)|||_0$. Clearly u is convex and coincides with ℓ up to the second order terms.

Proof of 12.1.E. Every variation of the path f_t can be written in the form $f_{t,\varepsilon} = h_{t,\varepsilon}^{-1} \circ f_t$, where $h_{t,\varepsilon}$ is a smooth family of loops with $h_{t,0} = \mathbb{1}$. The Hamiltonian $H(x,t,\varepsilon)$ of $\{h_{t,\varepsilon}\}$ belongs to $\mathcal{V}$. Further,

$$F(x,t,\varepsilon) = -H(h_{t,\varepsilon}x,t,\varepsilon) + F(h_{t,\varepsilon}x,t),$$

thus

$$\ell(\varepsilon) = |||F - H(\varepsilon)|||_0. \qquad \square$$

Solve the following exercises using 12.1.E.

Exercise 12.1.F. Assume that for every t the function $F(x,t)$ has a unique point of maximum and minimum respectively, and moreover these points are non-degenerate in the sense of Morse theory. Then *for every* variation of $\{f_t\}$ the length function $\ell(\varepsilon)$ is smooth. *Hint:* Use the implicit function theorem.

Exercise 12.1.G. Assume that the maximum set of F has non-empty interior. Construct a variation such that the corresponding length function is not smooth at 0.

12.2 Description of geodesics

We say that a function $F \in \mathcal{F}_0$ *has fixed extrema* if there exist two points $x_-, x_+ \in M$ such that $F(x_-, t) = \min_x F(x, t)$ and $F(x_+, t) = \max_x F(x, t)$ for all $t \in [0; 1]$, and moreover the function $F(x_+, t) - F(x_-, t)$ does not depend on t. The crucial point here is that the extremal points x_- and x_+ of $F(\cdot, t)$ are time-independent.

Theorem 12.2.A. *A path $\{f_t\}$ is a geodesic if and only if the corresponding Hamiltonian function $F \in \mathcal{F}_0$ has fixed extrema.*

In particular, every autonomous Hamiltonian flow is a geodesic. Hamiltonian functions with fixed extrema were introduced in [BP1] where they were called *quasi-autonomous*. Theorem 12.2.A above is essentially due to [LM2] (though the definition of a geodesic is given there in a slightly different way).

In order to prove Theorem 12.2.A we have to investigate in more details the structure of variations. Denote by $\mathcal{V}_1$ the tangent space to $\mathcal{H}_0$ at 0:

$$\mathcal{V}_1 = \{\frac{\partial H}{\partial \varepsilon}(0) | H \in \mathcal{V}\}.$$

Proposition 12.2.B. *The space $\mathcal{V}_1$ consists of all functions $G \in \mathcal{F}_0$ which satisfy*

$$\int_0^1 G(x, t)dt = 0$$

for all $x \in M$.

The proof of the proposition is absolutely analogous to 5.2.D and 6.1.C above where the case of time-periodic variations was settled.

Proof of Theorem 12.2.A. Let $F \in \mathcal{F}_0$ be a function such that $\|F(\cdot, t)\|$ does not depend on t. For a family $H(\varepsilon)$ from $\mathcal{V}$ put $G = H'(0) \in \mathcal{V}_1$. Define the functions $\ell(\varepsilon) = \||F - H(\varepsilon)|\|_0$ and $v(\varepsilon) = \||F - \varepsilon G|\|_0$. In view of 12.1.E the Hamiltonian F generates a geodesic if and only if $\ell(\varepsilon)$ has a critical point in the sense of 12.1.C at $\varepsilon = 0$ for every $H \in \mathcal{V}$. Since $\ell(\varepsilon)$ and $v(\varepsilon)$ coincide up to the second order terms, and $v(\varepsilon)$ is convex, this is equivalent to the fact that $v(\varepsilon)$ has a point of minimum at $\varepsilon = 0$ for every $G \in \mathcal{V}_1$. Therefore in order to prove Theorem 12.2.A it suffices to verify that the following conditions are equivalent:

(i) F has fixed extrema;

(ii) $\||F - \varepsilon G|\|_0 \geq \||F|\|_0$ for all $G \in \mathcal{V}_1$, $\varepsilon \in \mathbb{R}$.

Assume that (i) holds. Set $u = F - \varepsilon G$ for $G \in \mathcal{V}_1$. Then Proposition 12.2.B implies that

$$|||u|||_0 \geq \int_0^1 u(x_+, t) - u(x_-, t)dt = |||F|||_0,$$

and hence we get (ii).

Assume now that (ii) holds. Take

$$G(x, t) = F(x, t) - \int_0^1 F(x, t)dt.$$

The inequality (ii) yields

$$\max_x \int_0^1 F(x, t)dt - \min_x \int_0^1 F(x, t)dt \geq \int_0^1 \max_x F(x, t) - \min_x F(x, t)dt,$$

which is possible only if F has fixed extrema. $\qquad\square$

12.3 Stability and conjugate points

A geodesic is called *stable* if for every variation the length function $\ell(\varepsilon)$ attains its minimal value at 0. In other words, stable geodesics cannot be shortened by small variations with fixed end points. The problem of characterization of stable geodesics in full generality remains open.[1] Below we present a solution for a special class of *non-degenerate* geodesics. By definition, a geodesic is non-degenerate if for every t the corresponding Hamiltonian function has a unique point of maximum and minimum respectively, and moreover these points are non-degenerate in the sense of Morse theory. For instance, every autonomous path with unique non-degenerate maximum and minimum is a non-degenerate geodesic. Recall that for every variation of a non-degenerate geodesic the length function $\ell(\varepsilon)$ is smooth (see 12.1.F). The theory developed in this and the next two sections is due to Ustilovsky [U] (see also [LM2]).

Let $\{f_t\}$, $t \in [0; 1]$, $f_0 = \mathbb{1}$ be a non-degenerate geodesic generated by a Hamiltonian $F \in \mathcal{F}_0$. Denote by x_- and x_+ the time-independent points of minimum and maximum of $F(\cdot, t)$ respectively. Note that $x_\pm$ are fixed points of $\{f_t\}$ in this case. Consider the linearized flows f_{t*} on $T_{x_+} M$ and $T_{x_-} M$. We say that such a flow has a non-trivial T-periodic orbit with $T \neq 0$ if $f_{T*}\xi = \xi$ for some tangent vector $\xi \neq 0$.

---

[1] I believe that it can be solved by existing methods of non-smooth analysis. Note also that the first statement of Theorem 12.3.A below which provides a sufficient condition for stability holds true for *all* geodesics, see [LM3]. The proof involves theory of pseudo-holomorphic curves.

Theorem 12.3.A. *([U])*

- *Assume that the linearized flows have no non-trivial T-periodic orbits with $T \in (0, 1]$. Then $\{f_t\}$ is stable.*
- *Assume that $\{f_t\}$ is stable. Then the linearized flows have no non-trivial periodic orbits with $T \in (0, 1)$.*

This result can be interpreted as a description of conjugate points along geodesics of Hofer's metric. Namely, conjugate points correspond to non-trivial closed orbits of the linearized flow at $x_\pm$. Before the conjugate point the geodesic is stable and after the conjugate point it loses its stability (that is it can be shortened by a small variation). We refer to [U] for further information. The proof of 12.3.A is presented in 12.5 below. It is based on the second variation formula which we are going do describe in the next section.

Exercise 12.3.B. Conclude from 12.3.A that a *sufficiently short* segment of a non-degenerate geodesic is stable.

12.4 The second variation formula

Let $\{f_t\}$ be a non-degenerate geodesic generated by the Hamiltonian $F(x, t)$ with maximum/minimum points $x_\pm$. Denote by $C_\pm(t)$ the operator of the linearized equation at $x_\pm$ i.e.

$$\frac{d}{dt} f_{t*}(x_\pm) = C_\pm(t) f_{t*}(x_\pm).$$

Consider the spaces

$$V_\pm = \{\text{smooth maps } v : [0, 1] \to T_{x_\pm} M \mid v(0) = v(1) = 0\}.$$

Given a variation $\{f_{t,\varepsilon}\}$ of $\{f_t\}$, define the elements $v_\pm \in V_\pm$ by

$$v_\pm = \frac{d}{d\varepsilon}\bigg|_{\varepsilon=0} f_{t,\varepsilon} x_\pm.$$

It will be convenient to consider the maximum and the minimum part of the length function associated to the variation separately. Set

$$\ell_+(\varepsilon) = \int_0^1 \max_x F(x, t, \varepsilon)dt,$$

and

$$\ell_-(\varepsilon) = \int_0^1 \min_x F(x, t, \varepsilon)dt.$$

Clearly $\ell(\varepsilon) = \ell_+(\varepsilon) - \ell_-(\varepsilon)$.

Theorem 12.4.A. *([U])*

$$\frac{d^2\ell_\pm}{d\varepsilon^2}(0) = Q_\pm(v_\pm)$$

and hence

$$\frac{d^2\ell}{d\varepsilon^2}(0) = Q_+(v_+) - Q_-(v_-),$$

where

$$Q_\pm(v) = -\int_0^1 \Big(\Omega(C_\pm^{-1}\dot v, \dot v) + \Omega(\dot v, v)\Big)dt.$$

Example (Isoperimetric inequality). Consider the standard symplectic plane $\mathbb{R}^2(p,q)$ endowed with the symplectic form $\omega = dp \wedge dq$. Let $v : [0,1] \to \mathbb{R}^2$ be a smooth curve with $v(0) = v(1) = 0$. Recall the following notions from Euclidean geometry:

$$\text{length }(v) = \int_0^1 |\dot v|dt, \qquad \text{energy }(v) = \int_0^1 |\dot v|^2 dt,$$

$$\text{area }(v) = \int_0^1 \langle \frac{1}{2}(pdq - qdp), \dot v\rangle dt$$

$$= \frac{1}{2}\int_0^1 p\dot q - q\dot p \; dt$$

$$= \frac{1}{2}\int_0^1 \omega(v, \dot v)dt,$$

where $v(t) = (p(t), q(t))$. The isoperimetric inequality reads

$$4\pi \text{ area }(v) \le (\text{ length }(v))^2 \le \text{ energy }(v).$$

Going back to our symplectic setting, we assume that the Hamiltonian near x_- is given by $F(x) = \pi\lambda(p^2 + q^2)$ with $\lambda > 0$. So we get the Hamiltonian system

$$\begin{cases} \dot p = -2\pi\lambda q \\ \dot q = 2\pi\lambda p \end{cases}$$

Setting $z = p + iq$ we obtain the linear equation $\dot z = 2\pi\lambda iz$. In this case $C_-(t) = 2\pi\lambda i$ and $C_-^{-1}(t) = -\frac{i}{2\pi\lambda}$ and, recalling that $\omega(\xi, i\xi) = |\xi|^2$, we get

$$Q_-(v) = -\int_0^1 \Big(\omega(-\frac{i}{2\pi\lambda}\dot v, \dot v) + \omega(\dot v, v)\Big)dt$$

$$= -\frac{1}{2\pi\lambda}\int_0^1 |\dot v|^2 dt - \int_0^1 \omega(\dot v, v)dt$$

$$= -\frac{1}{2\pi\lambda} \text{ energy }(v) + 2 \text{ area }(v).$$

Equation $\dot{z} = 2\pi\lambda i z$ has the solution $z(t) = e^{2\pi\lambda i t}z(0)$ for $t \in [0,1]$ and thus it has no orbits of period 1 for $\lambda < 1$. Combining Theorems 12.3.A and 12.4.A we get that $Q_-(v) \leq 0$ for all plane curves v with $v(0) = v(1) = 0$. Therefore

$$4\pi\lambda \ \text{area} \ (v) \leq \ \text{energy} \ (v)$$

for all $\lambda < 1$. Letting λ go to 1 we get the isoperimetric inequality.

For the proof of 12.4.A we need the following auxiliary statement.

Lemma 12.4.B. *Let $\{f_{t,\varepsilon}\}$ be a variation of the path $\{f_t\}$, and let $F(x,t,\varepsilon)$ be the corresponding Hamiltonian function. Then*

$$\int_0^1 \frac{\partial F}{\partial \varepsilon}(f_{t,\varepsilon}x, t, \varepsilon)dt = 0.$$

Proof.
1) The formula above is valid for any variation of the constant loop, that is when $f_t = 1$ for all t. The proof is analogous to the proof of 6.1.C above.
2) Consider now the general case. Write $f_{t,\varepsilon} = f_t \circ h_{t,\varepsilon}$, where $h_{t,\varepsilon}$ is a variation of the constant loop. Denote by $H(x,t,\varepsilon)$ the Hamiltonian of $\{h_{t,\varepsilon}\}$. Then $F(x,t,\varepsilon) = F(x,t) + H(f_t^{-1}x,t,\varepsilon)$, and hence

$$\frac{\partial F}{\partial \varepsilon}(x,t,\varepsilon) = \frac{\partial H}{\partial \varepsilon}(f_t^{-1}x,t,\varepsilon).$$

Therefore

$$\frac{\partial F}{\partial \varepsilon}(f_{t,\varepsilon}x,t,\varepsilon) = \frac{\partial H}{\partial \varepsilon}(h_{t,\varepsilon}x,t,\varepsilon).$$

The required statement follows now from step 1. $\square$

Proof of 12.4.A. We will calculate the second derivative of $\ell_+(\varepsilon)$. The function $\ell_-(\varepsilon)$ is treated analogously. We work in the standard symplectic coordinates $x = (p,q)$ near x_+ and write $\xi \cdot \eta$ for the Euclidean scalar product. Let i denote the complex structure $(p,q) \mapsto (-q,p)$. In this terminology $\Omega(\xi, i\eta) = \xi \cdot \eta$ and the Hamiltonian system reads

$$\frac{d}{dt}f_t x = i\frac{\partial F}{\partial x}(f_t x, t).$$

Thus

$$C_+(t) = i\frac{\partial^2 F}{\partial x^2}(x_+, t).$$

To simplify the formulae later on, we will introduce the following notation:

$$a = \frac{\partial^2 F}{\partial x \partial \varepsilon}(x_+, t, 0)$$

$$b = \frac{\partial x_+}{\partial \varepsilon}(t, 0)$$

$$c = \frac{\partial^2 F}{\partial \varepsilon^2}(x_+, t, 0)$$

$$K = \frac{\partial^2 F}{\partial x^2}(x_+, t).$$

The implicit function theorem insures that $F(\cdot, t, \varepsilon)$ has the unique point of maximum $x_+(t, \varepsilon)$ which depends smoothly on t and ε. Since

$$\ell_+(\varepsilon) = \int_0^1 F(x_+(t, \varepsilon), t, \varepsilon)dt$$

we get that

$$\frac{d\ell_+}{d\varepsilon}(\varepsilon) = \int_0^1 \frac{\partial F}{\partial x}(x_+(t, \varepsilon), t, \varepsilon)\frac{\partial x_+}{\partial \varepsilon}(t, \varepsilon) + \frac{\partial F}{\partial \varepsilon}(x_+(t, \varepsilon), t, \varepsilon)dt.$$

Since $x_+(t, \varepsilon)$ is a critical point of $F(\cdot, t, \varepsilon)$ this expression simplifies to

$$\frac{d\ell_+}{d\varepsilon}(\varepsilon) = \int_0^1 \frac{\partial F}{\partial \varepsilon}(x_+(t, \varepsilon), t, \varepsilon)dt.$$

Differentiating again with respect to ε and setting $\varepsilon = 0$ we get

$$\frac{d^2\ell_+}{d\varepsilon^2}(0) = \int_0^1 (a \cdot b + c)dt. \qquad (12.4.\text{C})$$

Differentiation of the equation $\frac{\partial F}{\partial x}(x_+(t, \varepsilon), t, \varepsilon) = 0$ with respect to ε at $\varepsilon = 0$ yields

$$Kb + a = 0.$$

The non-degeneracy condition guarantees that we can invert K and thus

$$b = -K^{-1}a. \qquad (12.4.\text{D})$$

In view of Lemma 12.4.B

$$\int_0^1 \frac{\partial F}{\partial \varepsilon}(f_{t,\varepsilon}x, t, \varepsilon)dt = 0.$$

Differentiating this with respect to ε at $\varepsilon = 0$ and $x = x_+$ we get

$$\int_0^1 (a \cdot v_+ + c)dt = 0. \tag{12.4.E}$$

Consider the Hamilton equation

$$\frac{d}{dt} f_{t,\varepsilon} x = i \frac{\partial F}{\partial x}(f_{t,\varepsilon} x, t, \varepsilon).$$

Differentiate it with respect to ε at $\varepsilon = 0$ and $x = x_+$. We obtain

$$\dot{v}_+ = iK v_+ + ia$$

or equivalently

$$a = -i\dot{v}_+ - K\, v_+. \tag{12.4.F}$$

Taking 12.4.D and 12.4.F together, we have

$$b = -K^{-1}(-i\dot{v}_+ - Kv_+) = K^{-1}i\dot{v}_+ + v_+.$$

In view of 12.4.C this implies

$$\begin{aligned}
\frac{d^2\ell_+}{d\varepsilon^2}(0) &= \int_0^1 (a \cdot b + c)dt \\
&= \int_0^1 (a \cdot K^{-1}i\dot{v}_+ + a \cdot v_+ + c)dt \\
&\stackrel{12.4.E}{=} \int_0^1 a \cdot K^{-1}i\dot{v}_+ dt \\
&\stackrel{12.4.F}{=} -\int_0^1 (i\dot{v}_+ + Kv_+) \cdot K^{-1}i\dot{v}_+ dt \\
&\stackrel{*}{=} -\int_0^1 (K^{-1}i\dot{v}_+ \cdot i\dot{v}_+ + v_+ \cdot i\dot{v}_+)dt \\
&= -\int_0^1 (-C_+^{-1}\dot{v}_+ \cdot i\dot{v}_+ + v_+ \cdot i\dot{v}_+)dt \\
&= -\int_0^1 (\Omega(C_+^{-1}\dot{v}_+, \dot{v}_+) + \Omega(\dot{v}_+, v_+))dt.
\end{aligned}$$

The equality $(*)$ follows from the fact that K is symmetric. Furthermore we have used that $C_+^{-1} = -K^{-1}i$ and $\Omega(\xi, i\eta) = \xi \cdot \eta$ (or equivalently $-\Omega(\xi, \eta) = \xi \cdot i\eta$). $\square$

12.5 Analysis of the second variation formula

Proposition 12.5.A. *For all $a \in V_+$ and for all $b \in V_-$ there exists a variation $\{f_{t,\varepsilon}\}$ of $\{f_t\}$ such that $v_+ = a$ and $v_- = b$.*

Proof. Take a function $G \in \mathcal{F}_0$ such that

$$\int_0^1 G(x,t)dt = 0$$

for all $x \in M$ (that is $G \in \mathcal{V}_1$ in the notation of 12.2). This choice will be specified later on. Define the variation $\{h_{t,\varepsilon}\}$ of the constant loop $h_{t,0} = \mathbb{1}$ as follows: $h_{t,\varepsilon}$ is the time-ε-map of the Hamiltonian $\int_0^t G(x,s)ds$. Consider a variation $f_{t,\varepsilon} = f_t h_{t,\varepsilon}$ of $\{f_t\}$. We have

$$v_+(t) = \frac{d}{d\varepsilon}\Big|_{\varepsilon=0} f_{t,\varepsilon} x_+$$

$$= \frac{d}{d\varepsilon}\Big|_{\varepsilon=0} f_t h_{t,\varepsilon} x_+$$

$$= f_{t*} \frac{d}{d\varepsilon}\Big|_{\varepsilon=0} h_{t,\varepsilon} x_+.$$

Our explicit construction of $\{h_{t,\varepsilon}\}$ yields

$$\frac{d}{d\varepsilon}\Big|_{\varepsilon=0} h_{t,\varepsilon} x_+ = \int_0^t \operatorname{sgrad} G(x_+, s)ds.$$

Set

$$c(t) = \frac{d}{dt} f_{t*}^{-1} a(t).$$

Choose now $G \in \mathcal{V}_1$ such that in the standard symplectic coordinates near x_+

$$G(x,t) = \Omega(x - x_+, c(t)).$$

Since $a(0) = a(1) = 0$ by definition of V_+ this condition is compatible with the fact that $G \in \mathcal{V}_1$. The explicit expression for G yields

$$\operatorname{sgrad} G(x_+, t) = c(t) = \frac{d}{dt} f_{t*}^{-1} a(t).$$

Integrating this equality and taking into account that $a(0) = 0$ we get that

$$\int_0^t \operatorname{sgrad} G(x_+, s)ds = f_{t*}^{-1} a(t).$$

This implies that $v_+(t) = a(t)$. The same argument works for x_-. Since x_+ and x_- are disjoint, the variations we constructed do not interfere. This completes the proof. $\qquad\square$

Proof of Theorem 12.3.A. In view of 12.4.A and 12.5.A we come to the conclusion that $\{f_t\}$ is stable if and only if both functionals Q_+ and $-Q_-$ are non-negative. Classical calculus of variations provides us with precise conditions when this happens. In what follows we will focus on Q_+. The same arguments work for Q_-.

Consider the Lagrangian $\mathcal{L}(\dot{v}, v) = -\Omega(C_+^{-1}\dot{v}, \dot{v}) - \Omega(\dot{v}, v)$. Let us investigate the critical points of the functional

$$\int_0^1 \mathcal{L}(\dot{v}(t), v(t))dt$$

on V_+.

First of all, we claim that $\mathcal{L}$ satisfies the Legendre condition, that is $\frac{\partial^2 \mathcal{L}}{\partial \dot{v}^2}$ is positive definite. Indeed, $\mathcal{L}(\dot{v}, v) = -iC_+^{-1}\dot{v} \cdot \dot{v} - i\dot{v} \cdot v$ and therefore $\frac{\partial^2 \mathcal{L}}{\partial \dot{v}^2} = -2iC_+^{-1}$.

Since

$$C_+ = i\frac{\partial^2 F}{\partial x^2}(x_+) \quad \text{we get } -iC_+^{-1} = i\left(\frac{\partial^2 F}{\partial x^2}(x_+)\right)^{-1} i$$

and so

$$\left\langle i\left(\frac{\partial^2 F}{\partial x^2}(x_+)\right)^{-1} i\xi, \xi \right\rangle = -\left\langle \left(\frac{\partial^2 F}{\partial x^2}(x_+)\right)^{-1} i\xi, i\xi \right\rangle.$$

The point x_+ is a non-degenerate maximum of F. Thus the matrix $\frac{\partial^2 F}{\partial x^2}(x_+)$ is negative definite and so is $\left(\frac{\partial^2 F}{\partial x^2}(x_+)\right)^{-1}$. We deduce that $\frac{\partial^2 \mathcal{L}}{\partial \dot{v}^2}$ is positive definite. The claim follows.

Since $\mathcal{L}$ is quadratic with respect to v, the constant path $v_0(t) \equiv 0$ is an extremal. For quadratic functionals satisfying the Legendre condition, classical calculus of variations tells us the following. If v_0 is the minimum then for all $T \in [0, 1)$ the Euler-Lagrange equation (which coincides with the Jacobi equation) has no non-trivial solutions with $v(0) = v(T) = 0$. Vice versa, if there are no such solutions for $T \in [0, 1]$ then v_0 is the minimum. It remains to relate the Euler-Lagrange equation to periodic orbits of the linearized flow f_{t*} on $T_{x_+}M$. The Euler-Lagrange equation reads

$$\frac{d}{dt}\frac{\partial \mathcal{L}}{\partial \dot{v}} = \frac{\partial \mathcal{L}}{\partial v} \qquad \Longleftrightarrow$$

$$\frac{d}{dt}(-2iC_+^{-1}\dot{v} + iv) = -i\dot{v} \qquad \Longleftrightarrow$$

$$\frac{d}{dt}(-2iC_+^{-1}\dot{v}) = \frac{d}{dt}(-2iv).$$

If v is a solution then $C_+^{-1}\dot{v} = v + \text{const}$. Setting $w = v + \text{const}$ we have

$$\begin{cases} C_+^{-1}\dot{w} = w \\ w(0) = w(T). \end{cases}$$

But this means precisely that w is a T-periodic orbit of the linearized flow. This proves Theorem 12.3.A. $\qquad\qquad\square$

12.6 Length minimizing geodesics

Let $I \subset \mathbb{R}$ be an interval. Consider a geodesic $\{f_t\}, t \in I$ such that all its tangent vectors have unit speed:

$$\left\| \frac{d}{dt} f_t \right\| = 1$$

for all $t \in I$. Such a geodesic is called *locally strictly minimal* (cf. 8.2 above) if for every $t \in I$ there exists a neighbourhood U of t in I such that

$$\rho(f_a, f_b) = |a - b|$$

for all $a, b \in U$.

Every geodesic on a Riemannian manifold is locally strictly minimal. This is essentially due to the fact that the exponential map of the Levi-Civita connection is a diffeomorphism on a neighbourhood of zero. There is no analogue of such a fact in Hofer's geometry. Still various examples serve as a motivation for the following conjecture.

Conjecture 12.6.A. Every geodesic of Hofer's metric is locally strictly minimal.

This was proved first for the standard symplectic $\mathbb{R}^{2n}$ in [BP1]. Later this conjecture was confirmed in [LM2] for some other symplectic manifolds including cotangent bundles, closed oriented surfaces and $\mathbb{C}P^2$.

It is useful to relax the notion of local strict minimality in the following way. We say that a geodesic $\{f_t\}$, $t \in I$ is *locally minimal* if for every $t \in I$ there exists a neighbourhood U of t in I such that for every $a, b \in U$ with $a < b$ the geodesic segment $\{f_t\}$, $t \in [a, b]$ minimizes the length in the homotopy class of paths with fixed end points.

Conjecture 12.6.B. Every geodesic of Hofer's metric is locally minimal.

In finite-dimensional Riemannian geometry the local minimality yields the local strict minimality. In Hofer's geometry this is true under the following additional assumption. Denote by $S \subset [0; +\infty)$ the length spectrum of $\text{Ham}(M, \Omega)$ (see 7.3 above). Set $a = \inf(S \setminus \{0\})$. As usual the infimum of the empty set is assumed to be $+\infty$.

Proposition 12.6.C. *([LM2]) Assume that $a > 0$. Then every locally minimal geodesic is in fact locally strictly minimal.*

Proof. Let $\{f_t\}$ be a locally minimal geodesic. Assume without loss of generality that its Hamiltonian has unit norm at each time moment t. Thus there exists $\varepsilon \in (0; a/2)$ with the following property. For every path α on $\mathrm{Ham}(M, \Omega)$ which joins $\mathbb{1}$ with f_ε and which is homotopic to $\{f_t\}$, $t \in [0; \varepsilon]$ one has length $\alpha \geq \varepsilon$. It suffices to prove that $\rho(\mathbb{1}, f_\varepsilon) = \varepsilon$. Assume on the contrary that there exists a path β which joins $\mathbb{1}$ with f_ε such that length $\beta < \varepsilon$. Write γ for the path $\{f_t\}$, $t \in [0; \varepsilon]$. Consider a loop formed by γ and β. Its length is strictly less than a. Hence, by the definition of a, this loop can be homotoped to an arbitrarily short loop. But this implies that the path γ can be homotoped to a path γ' whose length is arbitrarily close to the length of β. We get a contradiction with the inequalities length $\gamma' \geq \varepsilon$ and $\varepsilon >$ length β. $\square$

In particular, if for a manifold (M, Ω) the quantity a is strictly positive then 12.6.B implies 12.6.A. The condition $a > 0$ is hard to verify. It holds for Liouville manifolds (see 7.3.B above), in the case when $\pi_1(\mathrm{Ham}(M, \Omega))$ is finite (for instance, for surfaces or $\mathbb{C}P^2$), and in a few other examples in dimension 4. No example with $a = 0$ is known.

Conjecture 12.6.B is confirmed in [LM2] for an interesting class of symplectic manifolds. Let us describe this class in more details. Assume that M is semi-monotone (cf. 11.3.A), that is there exists a constant $k \geq 0$ such that $c_1(A) = k([\Omega], A)$ for every spherical homology class $A \in H_2(M, \mathbb{Z})$. Further, when M is open, assume in addition that M is "geometrically bounded" at infinity. For instance any cotangent bundle T^*N endowed with the standard symplectic structure is geometrically bounded. For the precise definition see [AL]. Then 12.6.B holds.

Let us turn to a question which we started to discuss in 8.2 above. Given a geodesic, what can one say about the length of the time interval on which it is minimal? There exists a beautiful approach to this question based on the careful study of closed orbits of the corresponding Hamiltonian flow. It becomes especially transparent in the case of autonomous geodesics, that is one-parameter subgroups of $\mathrm{Ham}(M, \Omega)$. Let $\{f_t\}$ be a one-parameter subgroup of $\mathrm{Ham}(M, \Omega)$. The standard ODE technique yields that there exists a strictly positive number $\tau > 0$ with the following property (see [HZ, Section 5.7]). *Every closed orbit of the flow $\{f_t\}$ on the time interval $[0; \tau]$ is a fixed point of the flow.*

Conjecture 12.6.D. Every one-parameter subgroup is minimal on the interval $[0; \tau]$.

This phenomenon was discovered by Hofer in [H2] for the case of $\mathbb{R}^{2n}$. It could be considered as the global version of the conjugate points criterion 12.3.A above.[2] For generalizations to non-autonomous flows as well as to other symplectic manifolds we refer to [BP1], [Si1], [LM2], [Sch3] and [MS1]. Conjecture 12.6.D provides an interesting existence mechanism for non-trivial closed orbits of autonomous Hamiltonian flows. Indeed, assume that we know a priori that the autonomous flow $\{f_t\}$ is **not** minimal on some time interval. Such an information can come for instance from conjugate points, or from shortening procedures described in 8.2, 8.3 above. Then 12.6.D guarantees the existence of non-trivial closed orbits on that interval. We refer to [LM2] and [P8] for further discussion.

The study of length minimizing properties of geodesics led to the understanding of the following surprising feature of Hofer's geometry.

12.6.E. The C^1-flatness phenomenon, [BP1]. There exists a C^1-neighbourhood $\mathcal{E}$ of the identity in the group $\mathrm{Ham}(\mathbb{R}^{2n})$ and a C^2-neighbourhood $\mathcal{C}$ of 0 in its Lie algebra $\mathcal{A}$ such that $(\mathcal{E}, \rho)$ is isometric to $(\mathcal{C}, \| \ \|)$.

Some generalizations can be found in [LM2]. It is instructive to compare the C^1-flatness phenomenon with Sikorav's theorem 8.2.A which states that every one-parameter subgroup of $\mathrm{Ham}(\mathbb{R}^{2n})$ remains a bounded distance of the identity. This can be interpreted as a positive curvature type effect which intuitively contradicts the flatness! It is still unclear how to resolve this paradox. We refer to [BP1],[HZ] for the proof of the C^1-flatness phenomenon (see also [P8] for further discussion).

There are several different approaches to results on local minimality presented above. All of them are rather complicated. In the next chapter we present one of them which as we believe could be developed further in order to prove Conjecture 12.6.B in full generality. We will discuss the following simplest case only.

Theorem 12.6.F. *Let (M, Ω) be a closed symplectic manifold with $\pi_2(M) = 0$. Then every one-parameter subgroup of $\mathrm{Ham}(M, \Omega)$ generated by a generic time-independent Hamiltonian on M is locally minimal.*

The proof is sketched in 13.4 below.

---

[2]The discovery of conjugate points was in fact motivated by Hofer's result.

Chapter 13

A Visit to Floer Homology

In the present chapter we sketch a proof of Theorem 12.6.F which states that every one-parameter subgroup of $\mathrm{Ham}(M, \Omega)$ generated by a generic Hamiltonian function is locally minimal provided $\pi_2(M) = 0$. Our approach is based on Floer homology. The exposition is nor complete neither 100% rigorous. Its purpose is to sketch a very complicated and still developing machinery rather than to make a systematic introduction. We follow closely two papers by M. Schwarz [Sch2, Sch3].

13.1 Near the entrance

Floer homology is one of the most powerful tools of modern symplectic topology. Its creation was motivated by the following question: *what are invariants of Hamiltonian diffeomorphisms?* Let us describe the plan of the answer. We work on a connected closed symplectic manifold (M, Ω) assuming for simplicity that the manifold is aspherical: $\pi_2(M) = 0$. The role of this assumption will be clear in a few moments. Let us introduce some notations. Denote by $\widetilde{\mathrm{Ham}}(M, \Omega)$ the universal cover of the group of Hamiltonian diffeomorphisms with the base point $\mathbb{1}$. Write $\mathcal{L}M$ for the space of contractible loops $S^1 \to M$. Write $\mathcal{L}\,\mathrm{Ham}(M, \Omega)$ for the group of contractible loops $\{h_t\}$ of Hamiltonian diffeomorphisms based at $\mathbb{1}$ generated by functions from $\mathcal{H}$. To simplify the notations, we will often omit the dependence on M and Ω and write Ham for $\mathrm{Ham}(M, \Omega)$, etc.

The group $\mathcal{L}\,\mathrm{Ham}$ acts naturally on $\mathcal{L}M$ by the operators

$$T_h : z(t) \to h_t z(t).$$

The starting observation: there exists a natural map $\widetilde{\mathrm{Ham}} \to C^\infty(\mathcal{L}M)/\mathcal{L}\mathrm{Ham}$. Let us describe this map. Fix an element $\phi \in \widetilde{\mathrm{Ham}}$. Denote by $\mathcal{F}(\phi)$ the set of all Hamiltonians $F \in \mathcal{F}$ which generate ϕ. The group $\mathcal{L}\operatorname{Ham}$ acts transitively on $\mathcal{F}(\phi)$. This action is defined as follows. Take a loop $h \in \mathcal{L}\operatorname{Ham}$ and a Hamiltonian $F \in \mathcal{F}(\phi)$. Write $\{f_t\}$ for the Hamiltonian flow of F. Then $h(F)$ is defined as the normalized Hamiltonian function which generates the flow $h_t^{-1} \circ f_t$. Formula 1.4.D implies that

$$h(F)(x,t) = -H(h_t x, t) + F(h_t x, t).$$

Given $F \in \mathcal{F}(\phi)$, define a function $A_F : \mathcal{L}M \to \mathbb{R}$ called *the symplectic action*:

$$A_F(z) = \int_0^1 F(z(t), t)dt - \int_D \Omega,$$

where D is a disc in M bounded by the loop z. Since M is aspherical, the definition above does not depend on the choice of D.

Exercise 13.1.A. Prove that

$$(T_h^{-1})^* A_F = A_{h(F)}$$

for all $h \in \mathcal{L}\operatorname{Ham}$ and $F \in \mathcal{F}(\phi)$. *Hint:* Use that for every contractible loop $\{h_t\} \in \mathcal{L}\operatorname{Ham}$ generated by some $H \in \mathcal{H}$ the action vanishes identically on orbits: $A_H(\{h_t x\}) = 0$ for all $x \in M$. In fact on aspherical manifolds this is true even for non-contractible loops $\{h_t\}$. This difficult result was proved recently by Schwarz [Sch3].

This completes the description of the natural map which sends $\phi \in \widetilde{\operatorname{Ham}}$ to the equivalence class $[A_F] \in C^\infty(\mathcal{L}M)/\mathcal{L}\operatorname{Ham}$.

A function up to a diffeomorphism is a very rich object. For instance, in finite dimensions, a lot of information can be extracted from its critical points and the topology of the level sets. A powerful tool to get such information is Morse theory. In our situation, we have to deal with an infinite-dimensional manifold — the loop space $\mathcal{L}M$. The fundamental observation due to Floer is that there exists a suitable version of Morse theory which can be extended to this infinite-dimensional setting. We describe it in the next sections. Morse-Floer theory gives rise to a quite complicated structure associated canonically with $\widetilde{\operatorname{Ham}}$. Our main task is to figure out the place of Hofer's norm of ϕ inside this structure. As we will see, it is closely related to values of the action functionals at so-called homologically essential critical points.

We start with a digression to the finite dimensional situation.[1] The following remark will presumably help the reader to develop the right intuition. The manifold M can be naturally identified with the subset of $\mathcal{L}M$ consisting

---

[1]See the book [Sch1].

of constant loops. When the function $F \in \mathcal{F}$ is time-independent, the restriction of A_F to M is simply F. Thus the usual function theory on M "sits inside" the theory of the action functionals on $\mathcal{L}M$.

13.2 Morse homology in finite dimensions

Let F be a Morse function on a closed connected N-dimensional manifold M. We write $\mathrm{Crit}\, F$ for the set of critical points of F. Denote by $i(x)$ the Morse index[2] of a critical point x, and by $\mathrm{Crit}_m F$ the set of critical points with Morse index m. Denote by $C(F)$ the vector space over $\mathbb{Z}_2$ generated by $\mathrm{Crit}\, F$, and by $C_m(F)$ its subspace generated by critical points of index m.

Take a **generic**[3] Riemannian metric r on M and consider the negative gradient flow

$$\frac{du}{ds}(s) = -\nabla_r F(u(s)).$$

Pick two points $x_-, x_+ \in \mathrm{Crit}\, F$.

Fact 13.2.A. The space of orbits $u(s)$ of the gradient flow which satisfy $u(s) \to x_-$ as $s \to -\infty$ and $u(s) \to x_+$ as $s \to +\infty$ is a smooth manifold of dimension $i(x_-) - i(x_+)$.

Note that this space carries a natural free $\mathbb{R}$-action. Indeed, if $u(t)$ is a solution, then $u(t+\mathrm{const})$ is a solution as well. Thus when $i(x_-) - i(x_+) = 1$ the quotient space is a 0-dimensional manifold. In fact one can show that *it consists of a finite number of points.* Denote by $k_r(x_-, x_+) \in \mathbb{Z}_2$ the parity of this number. Define a linear operator

$$\partial_r : C_m(F) \to C_{m-1}(F)$$

as follows. For every $x \in \mathrm{Crit}_m(F)$ put

$$\partial_r x = \sum_{y \in \mathrm{Crit}_{m-1}(F)} k_r(x, y) y.$$

Fact 13.2.B. The operator ∂_r is a differential: $\partial_r^2 = 0$. Thus $(C(F), \partial_r)$ is a complex.

Fact 13.2.C. The m-th homology group $H_m(C(F), \partial_r)$ of this complex is isomorphic to the homology $H_m(M, \mathbb{Z}_2)$ of the manifold.

---

[2]This is the number of negative squares in the normal form of $d_x^2 F$.

[3]Here and below "generic" should be understood as in the last paragraph of 4.2: generic metric is an element of a residual subset in the space of all metrics, etc.

In particular, though these homology groups depend on additional parameters F and r all of them are mutually isomorphic. A remarkable fact is that the isomorphisms can be arranged to a canonical family as follows.

Consider the space of pairs (F, r) where F is a function and r is a Riemannian metric. Choose two generic pairs $\alpha = (F_0, r_0)$ and $\beta = (F_1, r_1)$. Choose a generic path (F_s, r_s), $s \in \mathbb{R}$ such that $(F_s, r_s) \equiv (F_0, r_0)$ for $s \leq 0$ and $(F_s, r_s) \equiv (F_1, r_1)$ for $s \geq 1$. Consider the equation

$$\frac{du}{ds}(s) = -\nabla_{r_s} F_s(u(s)). \tag{13.2.D}$$

Let $x_- \in \operatorname{Crit} F_0$ and $x_+ \in \operatorname{Crit} F_1$ be two critical points. As before, for generic choices the space of solutions $u(s)$ which satisfy $u(s) \to x_-$ as $s \to -\infty$ and $u(s) \to x_+$ as $s \to +\infty$ is a smooth manifold of dimension $i(x_-) - i(x_+)$. Further, when $i(x_-) = i(x_+)$ there is only a finite number of solutions.[4] Denote by $b(x_-, x_+) \in \mathbb{Z}_2$ the parity of this number. Define a linear operator $I^{\beta,\alpha} : C_*(F_0) \to C_*(F_1)$ by

$$I^{\beta,\alpha}(x) = \sum_{i(y)=i(x)} b(x, y) y.$$

Fact 13.2.E.

- Every operator $I^{\beta,\alpha}$ is a chain map and defines an isomorphism

$$I_*^{\beta,\alpha} : H_*(C(F_0), \partial_{r_0}) \to H_*(C(F_1), \partial_{r_1}).$$

- The operator $I_*^{\beta,\alpha}$ does not depend on the choice of the generic path (F_s, r_s).

- $I_*^{\alpha,\alpha} = \mathbb{1}$ and $I_*^{\gamma,\beta} \circ I_*^{\beta,\alpha} = I_*^{\gamma,\alpha}$ for generic α, β, γ.

We call a family of operators $\{I_*^{\alpha,\beta}\}$ which satisfies the last property *a canonical family*.

Definition 13.2.F. Let (C, ∂) be a complex over $\mathbb{Z}_2$ with a given basis $B = \{e_1, \ldots, e_k\}$. An element $e \in B$ is called *homologically essential* if for every ∂-invariant subspace

$$K \subset \operatorname{Span}(B \setminus \{e\})$$

the induced map

$$H_*(K, \partial) \to H_*(C, \partial)$$

is **not** surjective.

---

[4]The crucial difference between equation 13.2.D and the gradient flow is that its solution space does not admit an $\mathbb{R}_+$-action when the family (F_s, r_s) depends non-trivially on s.

The following statement plays a crucial role in our future considerations. Let F be a generic Morse function. Assume that $x_+ \in M$ is its unique point of absolute maximum. For a generic Riemannian metric r on M consider the complex $(C(F), \partial_r)$ with the basis $\operatorname{Crit} F$.

Proposition 13.2.G. *The point $x_+ \in \operatorname{Crit} F$ is homologically essential.*

Sketch of the proof: Since the function F decreases along the trajectories of the negative gradient flow, the space

$$Q = \operatorname{Span}(\operatorname{Crit} F \setminus \{x_+\})$$

is ∂_r-invariant. Moreover $H_N(Q, \partial_r)$ vanishes, where $N = \dim M$. This reflects the fact that the manifold $\{F < a\}$ is open for $a \leq \max F$ and thus carries no fundamental class. We conclude that for every ∂-invariant subspace of Q the image of its N-th homology group in $H_N(C(F), \partial_r)$ vanishes. But

$$H_N(C(F), \partial_r) = H_N(M, \mathbb{Z}_2) = \mathbb{Z}_2.$$

Hence x_+ is homologically essential. $\qquad\qquad\square$

This proposition yields the following important property of the coefficients $b(x, y)$ which count (mod 2) solutions of 13.2.D. Let F be a generic Morse function with the unique absolute maximum x_+. Consider a generic family (F_s, r_s), $s \in \mathbb{R}$ as in 13.2.D such that F_s equals some Morse function F_0 for all $s \leq 0$ and $F_s \equiv F$ for all $s \geq 1$.

Corollary 13.2.H. *There exists $x \in \operatorname{Crit} F_0$ such that $b(x, x_+) \neq 0$.*

Proof. Consider the operator

$$I : (C(F_0), \partial_{r_0}) \to (C(F), \partial_{r_1})$$

defined by our data. If all $b(x, x_+)$ vanish, then the image of I is contained in $\operatorname{Span}(\operatorname{Crit} F \setminus \{x_+\})$. This is a ∂_{r_1}-invariant subspace. Moreover, its homology coincides with $H(C(F), \partial_{r_1})$ since I_* is an isomorphism. We get a contradiction with the fact that x_+ is homologically essential. $\qquad\square$

13.3 Floer homology

The theory described in the previous section essentially generalizes to the infinite-dimensional situation. We replace the manifold M by its space of contractible loops $\mathcal{L}M$, and the function space $C^\infty(M)$ by the space of symplectic action functionals A_F where $F \in \mathcal{F}$.

Let us start with the description of critical points of A_F. Denote by $\mathcal{P}(F) \subset \mathcal{L}M$ the set of contractible 1-periodic orbits of the Hamiltonian flow $\{f_t\}$ generated by F.

Exercise 13.3.A. Prove that the critical points of A_F are exactly the elements of $\mathcal{P}(F)$. Moreover, if the time-1 map f_1 is *non-degenerate* in the sense that its graph is transversal to the diagonal in $M \times M$, then every critical point of A_F is non-degenerate.

Denote by $\mathcal{J}$ the space of all Ω-compatible almost complex structures on M. Choose an element $J \in C^\infty(S^1, \mathcal{J})$. Every such J defines a Riemannian metric on $\mathcal{L}M$ in the following way. The tangent space to $\mathcal{L}M$ at a loop $z \in \mathcal{L}M$ consists of vector fields on M along z. Given two such vector fields, say $\xi(t)$ and $\eta(t)$ define their scalar product by

$$\int_0^1 \Omega(\xi(t), J(t)\eta(t))dt.$$

Given $J \in \mathcal{J}$ we write ∇_J for the gradient with respect to the Riemannian metric $\Omega(\xi, J\eta)$ on M.

Exercise 13.3.B. Prove that the gradient of A_F with respect to the Riemannian metric on $\mathcal{L}M$ described above is given by

$$\mathrm{grad}A_F(u) = J\frac{du}{dt}(t) + \nabla_{J(t)}F(u(t), t).$$

Thus in order to define Morse complex of A_F one has to investigate the following problem.

Find a smooth map $u : \mathbb{R}(s) \times S^1(t) \to M$ such that

$$\frac{\partial u}{\partial s}(s, t) + J(t)\frac{\partial u}{\partial t}(s, t) + \nabla_{J(t)}F(u(s, t), t) = 0 \tag{13.3.C}$$

and $u(s, t) \to z_\pm(t)$ as $s \to \pm\infty$, where $z_\pm \in \mathcal{P}(F)$.

Denote by $\mathcal{M}_{F,J}(z_-, z_+)$ the space of solutions of 13.3.C. Floer established that it has a very nice structure which we are going to describe now. The crucial point is that 13.3.C is a Fredholm problem. In fact we see that up to the 0-order terms this is just the familiar Cauchy-Riemann equation. It turns out that for generic F and J the space $\mathcal{M}_{F,J}(z_-, z_+)$ is a smooth manifold with respect to a natural topology. Further, the dimension $d(z_-, z_+)$ of this manifold alters neither under homotopies of the Hamiltonian path $\{f_t\}$, $t \in [0; 1]$ with fixed end points, nor under changes of J. Thus this number depends only on the lift of the map $\{f_1\}$ to the universal cover of $\mathrm{Ham}(M, \Omega)$.

We arrived at one of the most surprising points of the whole theory. Notice that finite-dimensional intuition suggests that the number $d(z_-, z_+)$ is simply the difference of the Morse indices of A_F at critical points z_- and z_+. However it is very easy to see that these indices are infinite. Nevertheless their difference makes sense! The general formula for $d(z_-, z_+)$ is quite complicated — it requires the notion of the Conley-Zehnder index which we will not introduce in the book. However the situation simplifies drastically for the case when $F \in \mathcal{F}(g_\varepsilon)$, where $\{g_t\}$ is a given Hamiltonian flow generated by a normalized time-independent Morse function G, and $\varepsilon > 0$ is small enough. From now on we describe Floer homology for Hamiltonians $F \in \mathcal{F}(g_\varepsilon)$.

Exercise 13.3.D.

- Show that for $\varepsilon > 0$ small enough every ε-periodic orbit of the autonomous flow $\{g_t\}$ is a fixed point of the flow. Thus $\mathcal{P}(\varepsilon G)$ coincides with the set $\mathrm{Crit}\, G$ of the critical points of G.
- Prove that for $F \in \mathcal{F}(g_\varepsilon)$ every 1-periodic orbit of the flow $\{f_t\}$ is contractible. Moreover, the map

$$j : \mathcal{P}(F) \to \mathrm{Crit}\, G, \quad z(t) \to z(0)$$

establishes bijection between 1-periodic orbits of $\{f_t\}$ and critical points of G.

Denote by $i(x)$ the Morse index of a critical point $x \in \mathrm{Crit}\, G$.

Fact 13.3.E (The dimension formula). For a generic $F \in \mathcal{F}(g_\varepsilon)$, and for all $z_\pm \in \mathcal{P}(F)$ the following equality holds:

$$d(z_-, z_+) = i(z_-(0)) - i(z_+(0)).$$

Note that the solution space $\mathcal{M}_F(z_-, z_+)$ admits a natural $\mathbb{R}$-action by shifts $u(s, t) \to u(s+c, t)$ where $c \in \mathbb{R}$. The boundary conditions guarantee that this action is free. Assume in addition that $d(z_-, z_+) = 1$. Then the dimension formula yields that the quotient space $\mathcal{M}_{F,J}(z_-, z_+)/\mathbb{R}$ is a zero dimensional manifold. A version of Gromov's compactness theorem for solutions of 13.3.C states that this manifold is compact, hence consists of a finite number of points. Denote by $k_J(z_-, z_+) \in \mathbb{Z}_2$ the parity of this number.

We proceed further exactly as in the finite-dimensional situation. Recall that critical points of A_F are identified with $\mathrm{Crit}\, G$ via the map j, see 13.3.D. Set $C_m(A_F) = C_m(G)$ and $C(A_F) = C(G)$. Fix a loop $J(t)$ and define a linear operator $\partial_J : C(A_F) \to C(A_F)$ as follows. For an element $x \in \mathrm{Crit}_m G$ set

$$\partial_J(x) = \sum_{y \in \mathrm{Crit}_{m-1} G} k_J(j^{-1}x, j^{-1}y)y.$$

Fact 13.3.F. For generic F and J the operator ∂_J is a differential: $\partial_J^2 = 0$. Thus $(C(A_F), \partial_J)$ is a chain complex.

Example 13.3.G. Consider equation 13.3.C for $F = \varepsilon G$ and for a constant loop $J(t) \equiv J$. Every solution $u(s)$ of the gradient flow equation

$$\frac{du}{ds}(s) = -\varepsilon \nabla_J G(u(s))$$

solves 13.3.C. A fairly involved argument [HS, Lemma 7.1] shows that for $\varepsilon > 0$ small enough these are the only solutions of the equation provided G and J are generic. Therefore $(C(A_F), \partial_J)$ coincides with the usual Morse complex of the function F!

The following discussion emphasizes an important and delicate point in Floer theory. Let us try to prove the statement above for the case when $d(z_-, z_+) = 1$. We have to show that every solution of equation 13.3.C is time-independent. Since G is time-independent, the solution space $\mathcal{M}_{\varepsilon G, J}(z_-, z_+)$ admits a natural $\mathbb{R} \times S^1$-action $u(s, t) \to u(s+c_1, t+c_2)$ where $(c_1, c_2) \in \mathbb{R} \times S^1$. Assume that $u(s, t)$ is a solution which depends non-trivially on t. Then the $\mathbb{R} \times S^1$-action is free in a neighbourhood of u, and thus the dimension $d(z_-, z_+)$ is at least 2. This contradiction proves that u cannot depend on t. Unfortunately, such an argument contains a serious gap. Namely, *time-independent* functions form a "very thin" set in $\mathcal{F}$, and thus no general position type argument can guarantee a priori that the space of solutions $\mathcal{M}_{\varepsilon G, J}(z_-, z_+)$ is a manifold! We refer to [HS] for a complete proof.

Let $F_0 \in \mathcal{F}(g_\delta)$ and $F_1 \in \mathcal{F}(g_\varepsilon)$ be two generic functions. Consider a generic family (F_s, J_s), $s \in \mathbb{R}$ where $F_s \in \mathcal{F}$, $J_s \in C^\infty(S^1, \mathcal{J})$ and $(F_s, J_s) \equiv (F_0, J_0)$ for $s \leq 0$ and $(F_s, J_s) \equiv (F_1, J_1)$ for $s \geq 1$. Exactly as in the finite dimensional situation we shall define a canonical homomorphism

$$I : (C(A_{F_0}), \partial_{J_0}) \to (C(A_{F_1}), \partial_{J_1}).$$

The following problem is analogous to 13.2.D.

Find a smooth map $u : \mathbb{R}(s) \times S^1(t) \to M$ such that

$$\frac{\partial u}{\partial s}(s, t) + J_s(t)\frac{\partial u}{\partial t}(s, t) + \nabla_{J_s(t)} F_s(u(s, t), t) = 0 \qquad (13.3.\mathrm{H})$$

and $u(s, t) \to z_\pm(t)$ as $s \to \pm\infty$, where $z_- \in \mathcal{P}(F_0)$ and $z_+ \in \mathcal{P}(F_1)$.

The analysis of this equation is quite similar to the one we sketched above. In particular, if $i(z_-) = i(z_+)$ then under certain genericity assumptions the space of solutions is zero dimensional and compact. Thus it consists of a finite number of points. Denote by $b(z_-, z_+) \in \mathbb{Z}_2$ the parity of this number. In contrast to the case of 13.3.C the solution space of 13.3.H is not invariant under translations $u(s,t) \to u(s+c,t)$ provided (F_s, J_s) depends non-trivially on the s-variable. Define now a linear map I as follows. For $x \in \mathrm{Crit}_m G$ set

$$I(x) = \sum_{y \in \mathrm{Crit}_m G} b(j^{-1}x, j^{-1}y)y. \tag{13.3.I}$$

In complete analogy with 13.2.E one can show that I induces an isomorphism I_* on homology which does not depend on the choice of the generic path (F_s, J_s). Moreover, exactly as in 13.2 the family of maps I_* associated with different choices of parameters (F, J) can be arranged into a canonical family. In particular, 13.3.G and 13.2.C above imply that for generic F and J

$$H_*(C(A_F), \partial_J) = H_*(M, \mathbb{Z}_2).$$

13.4 An application to geodesics

Now we have the necessary tools which will enable us to prove Theorem 12.6.F. Let G be a normalized Morse function on M with unique absolute maximum point x_+ and unique absolute minimum point x_-. Write $\{g_t\}$ for its Hamiltonian flow. Take any Hamiltonian path $\{f_t\}$, $t \in [0;1]$ with $f_0 = 1\!\!1, f_1 = g_\varepsilon$ such that $\{f_t\}$ is homotopic to $\{g_{\varepsilon t}\}$, $t \in [0;1]$ with fixed end points. Let $F \in \mathcal{F}$ be its normalized Hamiltonian.

Proposition 13.4.A. *The following inequalities hold:*

$$\int_0^1 \max_x F(x,t) \geq \varepsilon \max G,$$

$$\int_0^1 \min_x F(x,t) \leq \varepsilon \min G.$$

Theorem 12.6.F immediately follows from this proposition.

For the proof of 13.4.A we start with the following auxiliary statement. Consider the loop $h_t = f_t \circ g_{\varepsilon t}^{-1}$. Note that $A_F = T_h^* A_{\varepsilon G}$ in view of 13.1.A. Fix a generic *time-independent* almost complex structure $J_0 \in \mathcal{J}$ on M, and consider the loop $J(t) = h_{t*}^{-1} J_0 h_{t*}$. Denote by r_0 and r Riemannian metrics on $\mathcal{L}M$ associated to J_0 and $J(t)$ respectively. Then $r = T_h^* r_0$. Thus with the help of T_h one can identify the complexes $(C(A_{\varepsilon G}), \partial_{J_0})$ and $(C(A_F), \partial_J)$. Moreover this identification preserves the basis $\mathrm{Crit}\, G$ of both complexes pointwise.

Lemma 13.4.B. *The point x_+ is homologically essential in $(C(A_F), \partial_J)$.*

Proof. Indeed, 13.3.G and 13.2.G yield that x_+ is homologically essential in $(C(A_{\varepsilon G}), \partial_{J_0})$. The result immediately follows from the identification above. $\square$

Proof of 13.4.A. Fix $\delta \in (0; \epsilon)$. Let $a(s)$, $s \in \mathbb{R}$ be a non-decreasing function such that $a(s) \equiv 0$ for $s \leq 0$ and $a(s) \equiv 1$ for $s \geq 1$. Set

$$F_s(x, t) = (1 - a(s))\delta G(x) + a(s)F(x).$$

Choose a generic path $J_s : S^1 \to \mathcal{J}$ such that $J_s(t) \equiv J_0$ for $s \leq 0$ and $J_s(t) \equiv J(t)$ for $s \geq 1$. Consider the homomorphism I defined in 13.3.I. Set $z_+(t) = f_t x_+ \in \mathcal{P}(F)$. Using that x_+ is homologically essential (see 13.4.B) and arguing exactly as 13.2.H we get that $b(z_-, z_+) \neq 0$ for some $z_- \in \mathcal{P}(\delta G)$. Hence there exists a solution $u(s, t)$ of the problem 13.3.H. Consider the *energy integral*

$$E = \int_{-\infty}^{+\infty} ds \int_0^1 dt \, \Omega\left(\frac{\partial u}{\partial s}(s, t), J_s(t)\frac{\partial u}{\partial s}(s, t)\right).$$

Using equation 13.3.H and the fact that $\operatorname{sgrad} F = J\nabla_J F$ for every $J \in \mathcal{J}$ we compute

$$\Omega\left(\frac{\partial u}{\partial s}, J_s(t)\frac{\partial u}{\partial s}\right) = \Omega\left(\frac{\partial u}{\partial s}, \frac{\partial u}{\partial t}\right) - dF_s\left(\frac{\partial u}{\partial s}\right). \tag{13.4.C}$$

Let D_- and D_+ be discs in M bounded by z_- and z_+ respectively. Since $\pi_2(M) = 0$, we get that

$$\int_{-\infty}^{+\infty} ds \int_0^1 dt \, \Omega\left(\frac{\partial u}{\partial s}, \frac{\partial u}{\partial t}\right) = \int_{D_+} \Omega - \int_{D_-} \Omega. \tag{13.4.D}$$

Further,

$$dF_s\left(\frac{\partial u}{\partial s}\right) = \frac{d}{ds}(F_s(u(s, t), t)) - \frac{\partial F_s}{\partial s}(u(s, t), t). \tag{13.4.E}$$

Integrating 13.4.C and substituting 13.4.D,E we get that

$$E = A_{\delta G}(z_-) - A_F(z_+) + Q,$$

where

$$Q = \int_{-\infty}^{+\infty} ds \int_0^1 dt \, \frac{da}{ds}(s)(F(u(s, t), t) - \delta G(u(s, t))).$$

Clearly

$$Q \leq \int_0^1 \max_x (F(x, t) - \delta G(x))dt,$$

$$A_F(z_+) = A_{\epsilon G}(x_+) = \epsilon \max G,$$

and $E \geq 0$. Moreover, $A_{\delta G}(z_-) \leq \delta \max G$ since all closed orbits of the flow $\{g_{\delta t}\}$ are simply critical points of G.

Therefore

$$\int_0^1 \max_x (F(x,t) - \delta G(x)) \geq (\varepsilon - \delta) \max G.$$

Since this is true for all $\delta > 0$ we get that

$$\int_0^1 \max_x (F(x,t) \geq \varepsilon \max G.$$

This proves the first inequality in 13.4.A. The second one can be proved analogously.
$\square$

13.5 Towards the exit

We start with a short summary of the material presented in this chapter. Let (M, Ω) be an aspherical symplectic manifold.

13.5.A. To every generic function $F \in \mathcal{F}$ one associates a vector space $C(A_F)$ over $\mathbb{Z}_2$. This space is endowed with a preferred basis $\mathcal{P}(F)$ consisting of all contractible periodic orbits of the corresponding Hamiltonian flow. Also, $C(A_F)$ has a $\mathbb{Z}$-grading in terms of the Conley-Zehnder index. We described this grading explicitly in terms of the Morse index in the simplest case when F belongs to $\mathcal{F}(g_\varepsilon)$.

13.5.B. A generic choice of a loop of compatible almost complex structures $J \in C^\infty(S^1, \mathcal{J})$ defines a differential $\partial_J : C_*(A_F) \to C_{*-1}(A_F)$. The homology of the complex $(C(A_F), \partial_J)$ is isomorphic to $H_*(M, \mathbb{Z}_2)$.

13.5.C. Given generic (F_0, J_0) and (F_1, J_1) and a generic path (F_s, J_s) interpolating between them, there exists a canonical chain map $I : (C(A_{F_0}), \partial_{J_0}) \to (C(A_{F_1}), \partial_{J_1})$. The map I induces an isomorphism I_* on homology which does not depend on the choice of the path. The isomorphisms I_* form a canonical family in the sense of 13.2.E.

13.5.D. Assume that F_0 and F_1 generate the same element of $\widetilde{\mathrm{Ham}}(M, \Omega)$. Then $F_0 = h(F_1)$ for a loop $h \in \mathcal{L}\,\mathrm{Ham}(M, \Omega)$ (see 13.1.A). Given $J_0(t) \in C^\infty(S^1, \mathcal{J})$, set $J_1(t) = h_{t*}^{-1} J_0(t) h_{t*}$. Then the map $T_h : \mathcal{L}M \to \mathcal{L}M$ introduced in 13.1 identifies $(C(A_{F_0}), \mathcal{P}(F_0), \partial_{J_0})$ with $(C(A_{F_1}), \mathcal{P}(F_1), \partial_{J_1})$. This identification is much stronger than the one described in 13.5.C – it works on the chain level while the I's induce an isomorphism on the homological level only.

The structure described in 13.5.A–13.5.D above is the simplest part of what is called Floer homology theory associated to a symplectic manifold. The next statement relates it to Hofer's norm.

13.5.E. Assume that for some $J \in C^\infty(S^1, \mathcal{J})$ a contractible periodic orbit $z \in \mathcal{P}(F)$ is homologically essential in $(C(A_F), \mathcal{P}(F), \partial_J)$. Then

$$A_F(z) \leq \int_0^1 \max_x F(x, t) dt.$$

The proof is absolutely analogous to the proof of 13.4.A presented in 13.4 above. Further, denote by $\phi \in \widetilde{\mathrm{Ham}}(M, \Omega)$ the element generated by F. In view of 13.5.D the property of an orbit $z \in \mathcal{P}(F)$ to be homologically essential for some J is an intrinsic property of the fixed point $z(0)$ of ϕ. The value $A_F(z)$ does not depend on the specific choice of $F \in \mathcal{F}(\phi)$. Thus inequality 13.5.E provides a tool for estimating the Hofer distance between $\mathbb{1}$ and ϕ on the universal cover $\widetilde{\mathrm{Ham}}(M, \Omega)$. It remains to develop a machinery which enables one to decide which orbits are homologically essential. In 13.4.B above we worked out the simplest situation when z corresponds to the absolute maximum of a small time-independent Hamiltonian function. In [Sch3] Schwarz used a more sophisticated argument which enabled him to prove the local minimality of a wide class of geodesics for aspherical symplectic manifolds. In fact, there is an important additional piece of data associated canonically to $(C(A_F), \mathcal{P}(F))$, namely *the canonical real filtration* of $C(A_F)$. For $a \in \mathbb{R}$ denote by C^a the subspace of $C(A_F)$ generated by those $z \in \mathcal{P}(F)$ which satisfy $A_F(z) \leq a$. Since the action functional decreases along the trajectories of its negative gradient flow, this subspace is ∂_J-invariant. Thus one can speak about the relative homology groups $H(C^a/C^b, \partial_J)$. It turns out that this homology carries a lot of interesting information about ϕ. Such a filtration was first systematically used by Viterbo [V1]. Further results on this filtration were obtained by Oh [O4] and Schwarz [Sch3].

Let me mention that there exist two natural directions of generalizing the theory sketched above. The first one is to extend it to symplectic manifolds with non-trivial π_2. For such manifolds the action functionals A_F are *multi-valued* functions $\mathcal{L}M \to \mathbb{R}$. In fact, their differentials dA_F are well defined closed 1-forms on $\mathcal{L}M$. Floer homology in this situation is designed as a generalization of Morse-Novikov homology of closed 1-forms (see [HS]). The second direction is to extend cohomological operations (like the cup-product) to Floer homology (see [PSS]).

Let me mention finally that Floer homology is a very complicated structure associated to $\widetilde{\mathrm{Ham}}(M, \Omega)$. An interesting task which is still far from being fulfilled is to develop an algebraic language suitable for a transparent description of this structure. We refer to [Fu] for first steps in this direction.

Chapter 14

Geometry of Non-Hamiltonian Diffeomorphisms

In this chapter we discuss interrelations between the groups $\mathrm{Ham}(M,\Omega)$ and $\mathrm{Symp}_0(M,\Omega)$, and explain the role played by non-Hamiltonian diffeomorphisms in Hofer's geometry.

14.1 The flux homomorphism

Let (M,Ω) be a closed symplectic manifold and let $\mathrm{Symp}_0(M,\Omega)$ be the identity component of the group of symplectomorphisms (see 1.4.C). Given a path $\{f_t\}$ of symplectomorphisms with $f_0 = \mathbb{1}$, consider a vector field ξ_t on M which generates this path. Since $\mathcal{L}_{\xi_t}\Omega = 0$ then $i_{\xi_t}\Omega = \lambda_t$ is a family of closed 1-forms. Let us emphasize that these forms are not necessarily exact. Let us consider a basic example which we started to discuss in the beginning of the book (see 1.4.C).

Example 14.1.A. Consider the 2-torus $(\mathbb{T}^2 = \mathbb{R}^2/\mathbb{Z}^2, dp \wedge dq)$ and the following system

$$\begin{cases} \dot{p} = 0 \\ \dot{q} = 1 \end{cases}$$

corresponding to the path of symplectomorphisms $f_t(p,q) = (p, q+t)$. This path is generated by an autonomous family of closed 1-forms $\lambda_t = dp$.

The first question we want to investigate is the following one. Take any $f \in \mathrm{Symp}_0(M,\Omega)$ (say given by a precise formula). *How do we decide whether f is Hamiltonian or not?* A powerful tool which will enable us to answer this question in the example above is the notion of the *flux* homomorphism introduced by Calabi and studied further by Banyaga [B1]. We will discuss it in this chapter. Note that in general one cannot conclude that f is not Hamiltonian when it is generated by a non-exact 1-form. For instance, in the above example $f_1 = \mathbb{1}$ is Hamiltonian.

Let us introduce the following useful notion. Let $\{f_t\}, t \in [0,1]$ be a loop of symplectomorphisms with $f_0 = f_1 = \mathbb{1}$. Let $\{\lambda_t\}$ be the family of closed 1-forms generating this loop.

Definition. The *flux* of a loop is given by $\mathrm{flux}(\{f_t\}) = \int_0^1 [\lambda_t] dt \in H^1(M, \mathbb{R})$.

Let us give a more geometric description of the flux. Let C be a 1-cycle on M. Define a 2-cycle $\partial[C] = \bigcup_t f_t(C)$ which is the image of C under the flow f_t. Note that ∂ is a linear map from $H_1(M, \mathbb{Z})$ to $H_2(M, \mathbb{Z})$.

Exercise. Show that $(\mathrm{flux}(\{f_t\}), [C]) = ([\Omega], \partial[C])$ for all $[C] \in H_1(M, \mathbb{Z})$.

In particular, this shows that $\mathrm{flux}(\{f_t\})$ depends only on the homotopy class of $\{f_t\}$ in $\pi_1(\mathrm{Symp}_0(M, \Omega))$, since the right-hand side persists under homotopies. Thus we get a homomorphism

$$\mathrm{flux} : \pi_1(\mathrm{Symp}_0(M, \Omega)) \to H^1(M, \mathbb{R}).$$

Definition. The image of the flux homomorphism $\Gamma \subset H^1(M, \mathbb{R})$ is called the *flux group*.

If γ is represented by a Hamiltonian loop then $\mathrm{flux}(\gamma) = 0$. A fundamental result due to Banyaga ([B1]) states that the converse is true. That is $\mathrm{flux}(\gamma) = 0$ implies that γ can be homotoped to a Hamiltonian loop.

It will be useful to extend the notion of flux to arbitrary (not necessarily closed) smooth paths of symplectomorphisms. Given a path $\{f_t\}$ generated by a family of closed 1-forms λ_t we put

$$\mathrm{flux}(\{f_t\}) = \int_0^1 [\lambda_t] dt \in H^1(M, \mathbb{R}).$$

Take a symplectomorphism $f \in \mathrm{Symp}_0(M, \Omega)$ and choose any path $\{f_t\}$ with $f_0 = \mathbb{1}$ and $f_1 = f$. Clearly, $\mathrm{flux}(\{f_t\})$ depends on the choice of the path connecting $\mathbb{1}$ with f, but the difference between the fluxes of any two such paths belongs to Γ. Thus we get a mapping

$$\Delta : \mathrm{Symp}_0(M, \Omega) \to H^1(M, \mathbb{R})/\Gamma.$$

We leave it as an exercise to check that Δ is a homomorphism. Banyaga showed that $\ker(\Delta) = \mathrm{Ham}(M, \Omega)$, thus

$$\mathrm{Symp}_0(M, \Omega)/\mathrm{Ham}(M, \Omega) = H^1(M, \mathbb{R})/\Gamma.$$

Exercise 14.1.B. Combine this with Theorem 1.5.A above and show that every normal subgroup of $\mathrm{Symp}_0(M,\Omega)$ has the form $\Delta^{-1}(K)$ for some subgroup K of $H^1(M,\mathbb{R})/\Gamma$.

Let us calculate the flux group for $(\mathbb{T}^2, dp \wedge dq)$ (see 14.1.A above). Make the following identifications:

$$H_1(\mathbb{T}^2, \mathbb{Z}) = \mathbb{Z}^2, \ H_2(\mathbb{T}^2, \mathbb{Z}) = \mathbb{Z}, \ H^1(\mathbb{T}^2, \mathbb{Z}) = \mathbb{Z}^2 \subset \mathbb{R}^2 = H^1(\mathbb{T}^2, \mathbb{R}).$$

We claim that with this language $\Gamma = \mathbb{Z}^2$. Indeed, for each $\gamma \in \pi_1(\mathrm{Symp}_0(\mathbb{T}^2))$ we have

$$(\mathrm{flux}(\gamma), a) = ([dp \wedge dq], \partial a).$$

Here ∂ is the functional which takes $\mathbb{Z}^2$ to $\mathbb{Z}$, the value of the class $[dp \wedge dq]$ on the generator of $H_2(\mathbb{T}^2, \mathbb{Z})$ equals 1, and $a \in H_1(\mathbb{T}^2, \mathbb{Z})$. Thus $\mathrm{flux}(\gamma) \in H^1(\mathbb{T}^2, \mathbb{Z})$ which implies that $\Gamma \subset \mathbb{Z}^2$. On the other hand, the fluxes of full rotations of the torus are given by

$$\mathrm{flux}\big((p,q) \mapsto (p, q + t)\big) = [dp] \,,$$
$$\mathrm{flux}\big((p,q) \mapsto (p + t, q)\big) = -[dq] \,,$$

and we conclude that $\mathbb{Z}^2 \subset \Gamma$. Therefore $\Gamma = \mathbb{Z}^2$. In particular, we learn that $f_T(p,q) = (p, q + T)$ is Hamiltonian precisely when $T \in \mathbb{Z}$. Indeed, $\Delta(f_T) = T[dp] \bmod \mathbb{Z}^2$.

14.2 The flux conjecture

In general, it is not so easy to calculate the flux group. The simple question whether Γ is discrete turns out to be an important one by the following reason. Recall (see 1.4.F above) that the flux conjecture[1] states that $\mathrm{Ham}(M,\Omega)$ is C^∞-closed in $\mathrm{Symp}_0(M,\Omega)$.

Theorem 14.2.A. *If Γ is discrete then the flux conjecture holds.*

Proof. Let ϕ_k be a sequence of Hamiltonian diffeomorphisms C^∞-converging to $\phi \in \mathrm{Symp}_0(M,\Omega)$. Fix a path of symplectomorphisms which joins ϕ with the identity. Write $\lambda \in H^1(M,\mathbb{R})$ for its flux. We claim that there exists a sequence of paths connecting ϕ_k to ϕ whose fluxes ε_k converge to 0 when k goes to $+\infty$. Assume the claim. Then (see Figure 13) for each k there exists a loop with flux $\lambda + \varepsilon_k \in \Gamma$. Since Γ is discrete we conclude that ε_k must equal

---

[1]This is the C^∞-flux conjecture. Note that it is equivalent to the C^1-flux conjecture. In contrast to this, the C^0-flux conjecture is a much stronger statement (see [LMP1] for discussion).

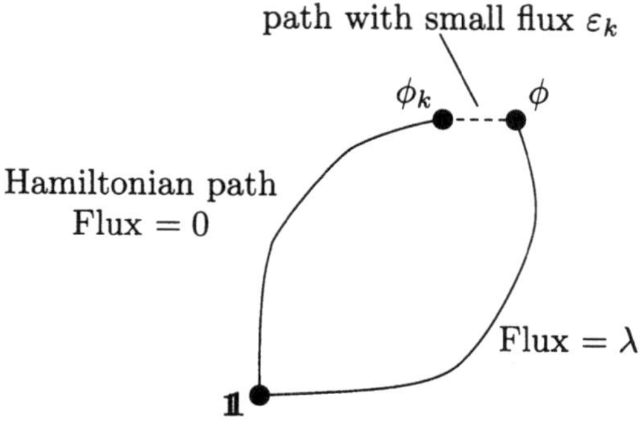

Figure 13

zero for large enough k. Thus ϕ is Hamiltonian since $\Delta(\phi) = 0 \bmod \Gamma$, so the theorem follows from the claim.

It remains to prove the claim. It suffices to find a sequence of paths γ_k connecting $\mathbb{1}$ with $\phi_k^{-1}\phi$ whose fluxes converge to 0.

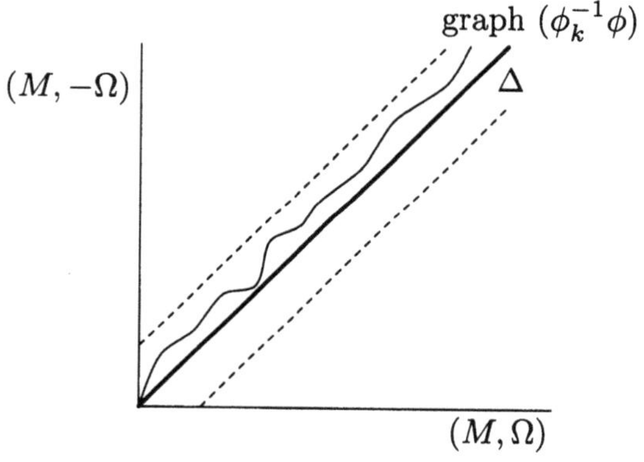

Figure 14

Note that $\phi_k^{-1}\phi \to \mathbb{1}$ when $k \to +\infty$, so graph $(\phi_k^{-1}\phi)$ is close to the diagonal L in $(M \times M, \Omega \oplus -\Omega)$. Now we use the following trick. Recall that L is a Lagrangian submanifold in $(M \times M, \Omega \oplus -\Omega)$.

Lemma. *Let (P, ω) be a symplectic manifold, and let $L \subset P$ be a closed Lagrangian submanifold. There exists a neighbourhood U of L in P and an embedding $f : U \to T^*L$ with the following properties:*

- *$f^*\omega_0 = \omega$, where ω_0 is the standard symplectic structure on T^*L;*
- *$f(x) = (x, 0)$ for all $x \in L$.*

This is a version of the Darboux theorem for Lagrangian submanifolds [MS].

Let us apply this lemma to the diagonal $L \subset M \times M$. It allows us to identify a tubular neighbourhood of L with a tubular neigbourhood of the zero section in T^*L. Since $T^*L \simeq T^*M$, we see that for large k the graph of $(\phi_k^{-1}\phi)$ corresponds to a section of T^*M that is to a closed 1-form α_k on M. Clearly $\alpha_k \to 0$ when $k \to +\infty$. Fix k large enough and consider the deformation of Lagrangian submanifolds

$$\text{graph } (s\alpha_k) \subset T^*M$$

for $s \in [0,1]$. For every s the submanifold graph $(s\alpha_k)$ is identified with the graph of a symplectomorphism of M. Denote this symplectomorphism by $\gamma_k(s)$. Clearly the flux of the path $\gamma_k(s)$ equals $[\alpha_k]$. This provides us with the required path, and hence completes the proof of the theorem. $\qquad\square$

In the proof above the main difficulty of the flux conjecture becomes especially transparent. It lies in the fact that for an arbitrary Hamiltonian isotopy $\{f_t\}$, the graph of $\{f_t\}$ may leave the tubular neighbourhood of L for some t. It will come back to this neighbourhood, but the question is whether it is then still the graph of an exact 1-form. Note also that the converse of the above theorem is also true (see [LMP1]).

Corollary. *If $[\Omega] \in H^2(M,\mathbb{Q})$ then the flux conjecture holds.*

Proof. We can rewrite the assumption as $[\Omega] \in \frac{1}{k}H^2(M,\mathbb{Z})$ for some $k \in \mathbb{Z}$. So

$$(\text{flux}(\gamma), [C]) = ([\Omega], \partial[C]) \in \frac{1}{k}\mathbb{Z}$$

for every $\gamma \in \pi_1(\text{Symp}_0(M,\Omega))$. Therefore $\Gamma \subset \frac{1}{k}H^1(M,\mathbb{Z}) \subset H^1(M,\mathbb{R})$ and we deduce that Γ is discrete. Now the above theorem implies that the flux conjecture holds. $\qquad\square$

14.3 Links to "hard" symplectic topology

Here is a more conceptual proof of the flux conjecture for $\mathbb{T}^2$. Clearly the C^k-closure of $\text{Ham}(M,\Omega)$ in $\text{Symp}_0(M,\Omega)$ is a normal subgroup of $\text{Symp}_0(M,\Omega)$. Hence in view of 14.1.B above it suffices to show that for every

$$\alpha \in H^1(\mathbb{T}^2,\mathbb{R})/\Gamma \setminus \{0\} \quad \textit{there exists} \quad f \in \text{Symp}_0(M,\Omega) \quad \textit{with} \quad \Delta(f) = \alpha$$

which cannot be represented as a limit of Hamiltonian diffeomorphisms. Without loss of generality take f to be a shift $(p,q) \mapsto (p, q+T)$, where $T \notin \mathbb{Z}$. The famous Arnold conjecture proved in [CZ] states that every $\phi \in \text{Ham}(T^2)$ has a fixed point. Thus if $\phi_k \to f$, this would imply that f has a fixed point too, which is a contradiction. This argument is due to M. Herman (1983), and it works for $\mathbb{T}^{2n}$ as well. Note also, that it proves the C^0-flux conjecture.

One can try to generalize this argument to other symplectic manifolds. The idea is to consider instead of the limiting behaviour of fixed points the limiting behaviour of Floer homology. In this way, one gets the following result.

Theorem 14.3.A. *([LMP1]) Assume that the first Chern class $c_1(TM)$ vanishes on $\pi_2(M)$. Then the flux conjecture holds.*

Here is another application of the theory of pseudo-holomorphic curves to the flux conjecture, which was found in [LMP2] for 4-dimensional symplectic manifolds, and later on proved in [McD2] in full generality.

Theorem 14.3.B. *The rank of the flux group Γ is finite and satisfies*

$$\mathrm{rank}_{\mathbb{Z}}\, \Gamma \leq b_1(M) = \dim H^1(M, \mathbb{R}).$$

As a corollary we get that for symplectic manifolds whose first Betti number equals 1, the group Γ is discrete, and therefore the flux conjecture holds.

14.4　Isometries in Hofer's geometry

No natural metric on $\mathrm{Symp}_0(M, \Omega)$ is known. However, one can include general non-Hamiltonian symplectomorphisms into the framework of Hofer's geometry as follows (see [LP]). Observe that the group $\mathrm{Symp}(M, \Omega)$ of all symplecto-morphisms of (M, Ω) acts on $\mathrm{Ham}(M, \Omega)$ by isometries. For $\phi \in \mathrm{Symp}(M, \Omega)$, define

$$T_\phi : \mathrm{Ham}(M, \Omega) \to \mathrm{Ham}(M, \Omega)\, , \quad f \mapsto \phi f \phi^{-1}.$$

One can easily check that T_ϕ is well defined and that it is an isometry with respect to Hofer's metric ρ.

Definition. An isometry T is *bounded* if $\sup \rho(f, Tf) < \infty$, where the supremum is taken over all $f \in \mathrm{Ham}(M, \Omega)$.

If ϕ is Hamiltonian, then the isometry T_ϕ is bounded. Indeed, the following inequality gives us a universal bound independent of f:

$$\rho(f, T_\phi f) = \rho(f, \phi f \phi^{-1}) = \rho(\mathbb{1}, \phi f \phi^{-1} f^{-1}) \leq 2\rho(\mathbb{1}, \phi).$$

Define the set $BI_0 \subset \mathrm{Symp}_0(M, \Omega)$ as the set of all $\phi \in \mathrm{Symp}_0(M, \Omega)$ such that T_ϕ is bounded. As we have seen above $\mathrm{Ham}(M, \Omega) \subset BI_0$.

Exercise 14.4.A. Prove that BI_0 is a normal subgroup of $\mathrm{Symp}_0(M, \Omega)$.

The next conjecture suggests a characterization of Hamiltonian diffeomor-phisms in metric terms.

Conjecture. $\mathrm{Ham}(M, \Omega) = BI_0$.

Theorem 14.4.B. *([LP]) The conjecture is true for surfaces of genus ≥ 1 and their products.*

We will give the idea of the proof for the case when M is the 2-torus $\mathbb{T}^2$. It follows from 14.4.A that in order to prove the theorem it suffices to show the following. Given

$$(a, b) \in H^1(\mathbb{T}^2; \mathbb{R}) \setminus \Gamma = \mathbb{R}^2 \setminus \mathbb{Z}^2,$$

there exists a symplectomorphism $\phi \in \mathrm{Symp}_0(\mathbb{T}^2)$ with $\Delta(\phi) = (a, b) \mod \mathbb{Z}^2$ such that the corresponding isometry T_ϕ is unbounded. Assume without loss of generality that $a = 0$, $b \in (0; 1)$ and $\phi(p, q) = (p, q + b)$. Note that for a curve $C = \{q = 0\}$ we have $C \cap \phi(C) = \emptyset$. Let $F = F(q)$ be a normalized Hamiltonian on $\mathbb{T}^2$ whose support lies in a small neigbourhood of C and such that $F|_C \equiv 1$. Denote by f_t the corresponding Hamiltonian flow, and consider the flow formed by the commutators $g_t = \phi^{-1} f_t^{-1} \phi f_t$. This flow is generated by the Hamiltonian $G(q) = F(q) - F(q + b)$. Since $G|_C \equiv 1$ we get from 7.4.A that $\rho(\mathbb{1}, g_t)$ goes to infinity when $t \to \infty$. Thus the isometry T_ϕ is unbounded.

The problem of describing all isometries of $(\mathrm{Ham}(M, \Omega), \rho)$ is open and seems to be very difficult even for surfaces. In connection to this let me recall the following classical result due to Mazur-Ulam [MU] (1932): every 0-preserving surjective isometry of a linear normed space is a linear map. It would be interesting to prove, or disprove, a non-linear version of this statement: every $\mathbb{1}$-preserving isometry of $\mathrm{Ham}(M, \Omega)$ is a group isomorphism (maybe after composition with the involution $f \mapsto f^{-1}$). Assume for a moment that this was indeed true. Then Banyaga's theorem 1.5.D would yield that each such isometry (up to the abovementioned involution) equals T_ϕ, where $\phi : M \to M$ is either a symplectomorphism, or an anti-symplectomorphism (that is $\phi^* \Omega = -\Omega$). This would mean that Hofer's geometry determines the symplectic topology.

Bibliography

[A]   M. Abreu *Topology of symplectomorphism groups of $S^2 \times S^2$*, Invent. Math. 131 (1998), 1–23.

[AM]   M. Abreu & D. McDuff *Topology of symplectomorphism groups of rational ruled surfaces*, Journal of the A.M.S. 13 (2000), 971–1009.

[AS]   L. Ahlfors & L. Sario *Riemann Surfaces*, Princeton University Press, 1960.

[Ar]   V. Arnold *Mathematical methods in Classical Mechanics*, Springer, Berlin, 1978.

[AK]   V. Arnold & B. Khesin *Topological methods in hydrodynamics* Springer, 1998.

[AL]   M. Audin & J. Lafontaine *Holomorphic curves in symplectic geometry*, Progress in mathematics 117, Birkhäuser, 1994.

[B1]   A. Banyaga *Sur la structure du groupe des difféomorphismes qui préservent une forme symplectique*, Comm. Math. Helv. 53 (1978), 174–227.

[B2]   A. Banyaga *The structure of classical diffeomorphism groups*, Mathematics and its applications 400, Kluwer Academic Publisher's Group, 1997.

[BP1]   M. Bialy & L. Polterovich *Geodesics of Hofer's metric on the group of Hamiltonian diffeomorphisms*, Duke Math. J. 76 (1994), 273–292.

[BP2]   M. Bialy & L. Polterovich *Invariant tori and symplectic topology*, Amer. Math. Soc. Transl. 171 (1996), 23–33.

[Ch]   Y. Chekanov *Lagrangian intersections, symplectic energy, and areas of holomorphic curves*, Duke Math. J. 95 (1998), 213–226.

[CFS]   I. Cornfeld, S. Fomin & Y. Sinai *Ergodic Theory*, Springer, 1982.

[CZ]   C. Conley & E. Zehnder *The Birkhoff-Lewis fixed point theorem and a conjecture of V.I. Arnold*, Invent. Math. 73 (1983), 33–49.

[DFN]   B.A. Dubrovin, A.T. Fomenko & S.P. Novikov *Modern Geometry – methods and applications, Part I*, Springer, 1984.

[EE]   C. Earle & J. Eells, *A fibre bundle description of Teichmüller theory*, J. Diff. Geometry 3 (1969), 19–43.

[EP]   Y. Eliashberg & L. Polterovich *Bi-invariant metrics on the group of Hamiltonian diffeomorphisms*, Internat. J. Math. 4 (1993), 727–738.

[F]   A. Floer *Morse theory for Lagrangian intersection theory*, J. Diff. Geometry 28 (1988), 513–517.

[Fu]   K. Fukaya *Morse homotopy and its quantization* Geometric topology (Athens, GA, 1993), 409–440, AMS/IP Stud. Adv. Math., 2. 1, Amer. Math. Soc., Providence, RI, 1997.

[GH]   P. Griffiths & J. Harris *Principles of algebraic geometry*, Wiley, New York, 1978.

[G1]   M. Gromov *Pseudoholomorphic curves in symplectic manifolds*, Invent. Math. 82 (1985), 307–347.

[G2]   M. Gromov *Positive curvature, macroscopic dimension, spectral gaps and higher signatures*, in "Functional Analysis on the eve of the 21st century", S. Gindikin, J. Lepowsky, R. Wilson eds., Birkhäuser, 1996.

[GLS] V. Guillemin, E. Lerman & S. Sternberg *Symplectic fibrations and multiplicity diagrams*, Cambridge University Press, 1996.

[He]   M. Herman *Inégalités à priori pour des tores lagrangiens invariants par des difféomorphismes symplectiques*, Publ. Math. IHES 70 (1989), 47–101.

[H1]   H. Hofer *On the topological properties of symplectic maps*, Proc. Royal Soc. Edinburgh 115A (1990), 25–38.

[H2]   H. Hofer *Estimates for the energy of a symplectic map*, Comm. Math. Helv. 68 (1993), 48–72.

[HS]   H. Hofer & D. Salamon *Floer Homology and Novikov rings*, in "Volume in Memory of Andreas Floer" (H. Hofer, C. Taubes, A. Weinstein, E. Zehnder, eds.), Progress in Mathematics, Birkhäuser, 1995, pp. 483–524.

[HZ]   H. Hofer & E. Zehnder *Symplectic invariants and Hamiltonian dynamics*, Birkhäuser Advanced Texts, Birkhäuser Verlag, 1994.

[Ki]   Y. Kifer *Random dynamics and its applications*, in "Proceedings of the ICM, Berlin, 1998", Vol. II, pp. 809–818.

[L]   F. Lalonde *Energy and capacities in symplectic topology*, AMS/IP Studies in Advanced Mathematics, Vol.2.1 (1997), 328–374

[LM1] F. Lalonde & D. McDuff *The geometry of symplectic energy*, Ann. of Math. 141 (1995), 349–371.

[LM2] F. Lalonde & D. McDuff *Hofer's L^∞-geometry: energy and stability of Hamiltonian flows I, II*, Invent. Math 122 (1995), 1–33, 35–69.

[LM3] F. Lalonde & D. McDuff *Local non-squeezing theorems and stability,* Geom. and Funct. Anal. 5 (1995), 364–386.

[LMP1] F. Lalonde, D. McDuff & L. Polterovich *On the flux conjectures,* "Geometry, Topology and Dynamics, ed F. Lalonde, Proceedings of the CRM 1995 Workshop in Montreal, the CRM Special Series of the AMS". Vol. 15, 1998, pp. 69–85.

[LMP2] F. Lalonde, D. McDuff & L. Polterovich *Topological rigidity of Hamiltonian loops and quantum homology,* Invent. Math. 135 (1999), 69–85.

[LP] F. Lalonde & L. Polterovich *Symplectic diffeomorphisms as isometries of Hofer's norm,* Topology 36 (1997), 711–727.

[LS] F. Laudenbach & J. -C. Sikorav *Hamiltonian disjunction and limits of Lagrangian submanifolds,* Internat. Math. Research Notices 1994, no. 4 161–168.

[Mac] R. MacKay *A criterion for nonexistence of invariant tori for Hamiltonian systems,* Phys. D 36 (1989), 64–82.

[MU] S. Mazur & S.M. Ulam *Sur les transformations isometriques d'espaces vectoriels normes* C. R. Acad. Sci. Paris 194 (1932), 946–948.

[MS] D. McDuff & D. Salamon *Introduction to symplectic topology,* Oxford Mathematical Monographs, Oxford University Press, 1995.

[MSl] D. McDuff & J. Slimowitz *Hofer-Zehnder capacity and length minimizing Hamiltonian paths,* Preprint 2000.

[McD1] D. McDuff, *Rational and ruled symplectic 4-manifolds* Journal of the A.M.S. 3 (1990) 679–712; erratum 5 (1992), 987–988.

[McD2] D. McDuff *Quantum homology of fibrations over S^2,* Internat. J. Math. 11 (2000), 665–721.

[NN] A. Newlander & L. Nirenberg *Complex analytic coordinates in almost complex manifolds,* Ann. of Math. 65 (1957), 391–404.

[O1] Y.-G. Oh *Floer cohomology of Lagrangian intersections and pseudo-holomorphic disks I, II,* Comm. Pure Appl. Math. 46 (1993), 949–993, 995–1012.

[O2] Y.-G. Oh *Relative Floer and quantum cohomology and the symplectic topology of Lagrangian submanifolds,* in *Contact and symplectic geometry,* C.B. Thomas ed., Cambridge University Press, 1996, pp. 201–267.

[O3] Y.-G. Oh *Gromov-Floer theory and disjunction energy of compact Lagrangian embeddings,* Math. Res. Lett. 4 (1997), 895–905.

[O4] Y.-G. Oh *Symplectic topology as the geometry of action functional – I, II,* J. Diff. Geometry 46 (1997), 499–577 and Comm. in Analysis and Geometry 7 (1999), 1–55.

[P1]   L. Polterovich *Symplectic displacement energy for Lagrangian submanifolds*, Ergod. Th. & Dynam. Sys. 13 (1993), 357–367.

[P2]   L. Polterovich *An obstacle to non-Lagrangian intersections*, In "Volume in Memory of Andreas Floer" (H. Hofer, C. Taubes, A. Weinstein, E. Zehnder, eds.), Progress in Mathematics, Birkhäuser, 1995, pp. 575–586.

[P3]   L. Polterovich *Gromov's K-area and symplectic rigidity*, Geom. and Funct. Anal. 6 (1996), 726–739.

[P4]   L. Polterovich *Symplectic aspects of the first eigenvalue*, J. reine angew. Math. (Crelle's journal) 502 (1998), 1–17.

[P5]   L. Polterovich *Hofer's diameter and Lagrangian intersections*, Internat. Math. Research Notices 4 (1998), 217–223.

[P6]   L. Polterovich *Hamiltonian loops and Arnold's principle*, Amer. Math. Soc. Transl. (2), 180 (1997), 181–187.

[P7]   L. Polterovich *Precise measurements in symplectic topology*, in "Proceedings of the 2nd European congress of Mathematics", Vol. 2, Birkhäuser, 1998, pp. 159–166.

[P8]   L. Polterovich *Geometry on the group of Hamiltonian diffeomorphisms*, in "Proceedings of the ICM, Berlin, 1998", Vol. II, pp. 401–410.

[P9]   L. Polterovich *Hamiltonian loops from the ergodic point of view*, Journal of the European Math. Soc. 1 (1999), 87–107.

[PS]   L. Polterovich & K.F. Siburg *On the asymptotic geometry of area-preserving maps*, Math. Research Letters 7 (2000), 233–243.

[PSS]   S. Piunikhin, D. Salamon & M. Schwarz *Symplectic Floer-Donaldson theory and quantum cohomology*, In "Contact and symplectic geometry (Cambridge, 1994)", C.B. Thomas ed., Publ. Newton Inst., 8, Cambridge Univ. Press, Cambridge, 1996, pp. 171–200.

[Sch1] M. Schwarz *Morse homology*, Birkhäuser, 1993.

[Sch2] M. Schwarz *Introduction to symplectic Floer homology*, In "Contact and symplectic geometry (Cambridge, 1994)", C.B. Thomas ed., Publ. Newton Inst., 8, Cambridge Univ. Press, Cambridge, 1996, pp. 151–170.

[Sch3] M. Schwarz *On the action spectrum for closed symplectically aspherical manifolds*, Pacific J. of Math. 193 (2000), 419–461.

[Se]   P. Seidel π_1 *of symplectic automorphism groups and invertibles in quantum homology rings*, Geom. and Funct. Anal. 7 (1997), 1046–1095.

[S1]   J.-C. Sikorav *Quelques propriétés des plongements lagrangiens*, Mém. Soc. Math. France (N.S.) 46 (1991), 151–167.

[S2]   J.-C. Sikorav *Systèmes Hamiltoniens et topologie symplectique*, ETS EDITRICE PISA, 1990.

[Si1]   K.F. Siburg *New minimal geodesics in the group of symplectic diffeomorphisms*, Calc. Var. 3 (1995), 299–309.

[Si2]   K.F. Siburg *Action-minimizing measures and the geometry of the Hamiltonian diffeomorphism group*, Duke Math. J. 92 (1998), 295–319

[Sm]   S. Smale *An infinite dimensional version of Sard's theorem*, Am. J. Math. 87 (1973), 213–221.

[U]   I. Ustilovsky *Conjugate points on geodesics of Hofer's metric*, Diff. Geometry Appl. 6 (1996), 327–342.

[V1]   C. Viterbo *Symplectic geometry as the geometry of generating functions*, Math. Annalen 292 (1992), 685–710.

[V2]   C. Viterbo *Metric and isoperimetric problems in symplectic geometry*, Journal of the A.M.S. 13 (2000), 411–431.

Index

List of Symbols

$\| \ \|$, 20
$\||F\||$, 37
$\||F\||_0$, 91

$\mathcal{A}$, 5
$\mathcal{A}(M)$, 5
$\mathrm{Ad}_f\, G$, 9

c_1, 78
$c(A)$, 25

$e(A)$, 16
$E(F)$, 58

$\mathcal{F}$, 37
$\mathcal{F}_0$, 91
$\{F, G\}$, 9
flux, 118

$\gamma(L)$, 24.

$\mathcal{H}$, 37
$\mathcal{H}_0$, 91
$\mathrm{Ham}(M, \Omega)$, 6

$\ell(\varepsilon)$, 89
λ_L, 23
$\mathrm{length}\{f_t\}$, 20
$l(F)$, 41

$\mu(F)$, 59

$\nu(\gamma)$, 54
$\nu_+(\gamma)$, 65

ρ, 20

$\mathrm{sgrad}\, F$, 4
supp, 3
$\mathrm{Symp}(M, \Omega)$, 7
$\mathrm{Symp}_0(M, \Omega)$, 7

$\mathcal{V}$, 91

GPSR Compliance
The European Union's (EU) General Product Safety Regulation (GPSR) is a set
of rules that requires consumer products to be safe and our obligations to
ensure this.

If you have any concerns about our products, you can contact us on

ProductSafety@springernature.com

In case Publisher is established outside the EU, the EU authorized
representative is:

Springer Nature Customer Service Center GmbH
Europaplatz 3
69115 Heidelberg, Germany